ANCESTRAL GENOMICS

ANCESTRAL GENOMICS

AFRICAN AMERICAN HEALTH IN THE AGE OF PRECISION MEDICINE

Constance B. Hilliard

HARVARD UNIVERSITY PRESS
Cambridge, Massachusetts
London, England
2024

Printed in the United States of America

First printing

Library of Congress Cataloging-in-Publication Data
Names: Hilliard, Constance B., author.
Title: Ancestral genomics : African American health in the age of precision medicine / Constance B. Hilliard.
Description: Cambridge, Massachusetts ; London, England : Harvard University Press, 2024. | Includes bibliographical references and index.
Identifiers: LCCN 2023036126 | ISBN 9780674268609 (cloth)
Subjects: LCSH: African Americans—Medical care. | African Americans—Health and hygiene. | Comparative genomics. | Genomics—United States. | Social medicine—United States.
Classification: LCC RA448.5.B53 H55 2024 | DDC 362.1089/96073—dc23/eng/20230824
LC record available at https://lccn.loc.gov/2023036126

To my son, Kenneth José Tripp

CONTENTS

Prologue 1

1 Our History: In the Shadow of Timbuktu 13

2 The System: American Medicine's Incomplete Paradigm 35

3 Our Health: Tracking Down Ancestral Clues 81

4 The Algorithm: Applying Genetic Ancestry to Case Studies 111

5 Ancestral Genomics: Embracing a Multigenomic Nation 130

Epilogue 158

Notes *165*

Acknowledgments *185*

Index *189*

ANCESTRAL GENOMICS

PROLOGUE

STATIC CRACKLED through the loudspeaker in the waiting room. A female voice pronounced a name that reminded me vaguely of my own but for it being stretched and modulated with a few extra beats. During that year spent in Japan as a Fulbright Visiting Professor, I had adjusted to many things, including the staccato qualities of English words rendered into *Romaji* and a standard of basic literacy whereby even schoolchildren mastered two different alphabets in addition to Chinese characters. I stood up and followed a tall, young doctor into a windowless office.

"Mrs. Hilliard, your chronic joint pain is arthritis. There is a new medication on the market, but since you are headed back home to Texas this month, I would rather not start you on a new drug regimen," he said. I nodded.

He then abruptly asked, "What type of problems have you had with your kidneys in the past?"

"My kidneys?" I stuttered in confusion. "I don't think I've ever had problems with my kidneys." The doctor settled into a chair behind the desk and gave me a puzzled look.

"You appear to be suffering from kidney failure," he said, almost hesitantly.

My mouth dropped open. "Excuse me?" I said, struggling to catch my breath. A medical check-up to find the source of a growing pain in my hip had widened into a comprehensive medical exam, including blood tests. And now I had kidney failure?

He continued, "Given your lab results, quite frankly I have never seen such a healthy-looking patient at such a late stage of renal disease." My shoulders slumped in quiet despair while the sudden gravity under my feet dragged them in a slow shuffle to the door.

I returned home to the Dallas area later that month and braced myself for more bad news. However, the results of new lab tests showed something completely different.

"Your kidneys are perfectly healthy," my primary care physician announced.

He handed me a lab report and directed my attention to the left-hand corner of the sheet. A tiny box that read "Race—African American" had been checked. I gave him a puzzled look. He merely shrugged.

"As for your joint pain, your Japanese doctor was right. It's osteoarthritis. Maybe you'll need a hip replacement in a few years. But for the time being, just do more walking."

My mind raced with questions that I was too embarrassed to ask. First and foremost: why did a checked box for race provide an accurate diagnosis in an American lab if races were not biologically real? The entrance of a nurse carrying fresh paper gowns signaled that my visit was over.

Now that I knew my kidneys were functioning properly, the tension in my body eased. But new questions arose. Why did the physician in Japan believe I had renal failure? Why did the fact that I was African American affect the medical test results? For years my lectures and classroom manner had conveyed to students in my African history course that race was not biologically real. It was artificial—a social construct concocted from a human need to create social pecking orders. As early as 1950, much of the scientific community had come to a consensus that racial divisions were devised by humans rather than nature and that we were a single species that originated in Africa.[1] Why, then, was race suddenly a factor in my medical diagnosis?

Over the course of the next several months, I researched medical journals seeking racial and ethnic disparities in disease risk. Rather than easy answers, I found even more questions. Dozens of studies in medical journals and articles in the national media described a crisis of kidney failure among African Americans. Black people were dying at rates three to four times that of other American demographic groups, and controversy swirled around the issue of equitable allocations of the scant number of kidney organs available for transplantation. My community's high risk of precisely the disease with which I had been misdiagnosed left me more confused than ever. And why had my primary care physician in demographically diverse America been aware of something about my kidney function as a Black American that the Asian doctor had overlooked? Those questions launched what became the most iconoclastic journey of my life.

I soon learned that nephrologists in the United States, to avoid misdiagnosing healthy African Americans as my Japanese doctor had in my case, adjusted the "normal" range of the chemical waste product known as creatinine in individuals who self-identified as African American. It was mistakenly assumed that race was the operational factor—a paradigmatic error that would lead the US medical community away from clues that could ultimately prove lifesaving. In actuality, two individuals who were identified as Black because their ancestors emanated from West Africa would have dramatically different kidney readings if one hailed from the salt-rich coast and the other was genetically adapted to the sodium-deficient interior of the African continent. But this was a narrative that could only be elucidated if a transdisciplinary approach involving historical and ecological knowledge was applied to the medical science—and that would come only later.

In probing the epistemologies (that is, from the vantage of my historian's perch, ways of knowing used by geneticists, molecular biologists, and medical science in general), I learned something new about the very laws of nature—they are rhythmic. They dance, weaving and twerking to their own cross-rhythms. However, it is choreography of both sublimity and precision. All that fancy footwork leaves behind clues and traces that have a habit of turning up not merely in the electron microscopes of scientists but in the lived spaces of humans. I followed these traces and eventually stumbled upon both mysteries and hidden signposts that the medical world had overlooked in its foggy knowledge of non-European ancestral environments. I believe that these new insights into certain drivers behind my community's unusually high risk of hypertension, kidney failure, certain lethal cancers, and childbirth fatalities will turn out to be beneficial in immeasurable ways.

MY FINDINGS

I am an evolutionary historian in search of clues regarding the ways that our human migratory past and adaptation to new ecologies have impacted human health. But beneath it all lay the curiosity of an African historiographer, a translator of manuscripts, and an investigator into factors that influence different perspectives on verifying what is real and true. The early years of my career were spent deciphering handwritten texts from the vast medieval empires of West Africa's interior, Timbuktu in particular. The enormous wealth in this region was based on a gold-salt trade whose networks extended through the Sahara Desert and crossed the Mediterranean into Europe. The source of this African gold, imported into Europe in the Middle Ages, was hidden from foreign prowlers. The thirteenth-century ruler Mansa Musa of Mali, believed by Europeans to possess the largest gold reserves in the world, in fact did not. (He was actually just a middleman in a trade that reached beyond the imperial boundaries of ancient

Mali into the decentralized societies to the west of interior Africa's merchant empires.)

However, as we look today at the lives of African Americans whose ancestors inhabited the regions surrounding these gold fields, we must acknowledge an awkward and stunning truth: it is not the gold that offers insights into Black health but rather the commodity that was so desperately scarce in the African interior that its elites readily traded their gold to obtain it—salt.

In the months after my return from Japan, I increasingly made connections between Timbuktu's history and my kidney disease misdiagnosis. American enslaved persons/farmers had come from the lands to the west of the wealthy merchant empires of Ghana, Mali, Songhay, and Timbuktu. These stateless societies were a thousand miles or so from the Atlantic coast.

The mere act of tapping the keystrokes on this page that record this distance in miles makes my stomach churn. For this was the march my ancestors were forced to make, in chains and the sweltering heat of the tropics after being kidnapped from the tiny villages that dotted that landscape. These societies were vulnerable to slavers not only because they lacked the military infrastructure of states but because their lands were perceived as worthless even by the standards of their African neighbors (because this geographic region lacked even the most minimal deposits of sodium that most humans need to survive). At the same time, this region was too far removed from the Atlantic coast to reap the benefits of sea salt. Its inhabitants had subsisted on 200 milligrams (mg.) of sodium a day while coastal West Africans, Europeans, and later Americans consumed between 3,400 and 5,000 mg. of sodium a day. As I delved even deeper into this matter of human populations' adapting to vastly different levels of sodium intake, it became clear that the medical literature made no distinctions between the ecological environments to which different human groups were biologically acclimated. In fact, the American practice of lumping populations into crude and unscientific racial categories almost ensured that the more medically relevant but nuanced details of human genetic adaptation would be missed.

For me, as an African American, racial terminology honors the truth of my ancestry, and, consequently, I take pride in being identified as such. However, that is in the context of culture and sociological research and federal forms that reflect opportunities for me to compete on a level playing field. It was my heroes of the civil rights movement who fought and died to gain recognition of the need for redress from the traumas of slavery. But racial designations are not science; when used in medical research, they tend to confuse rather than clarify our understanding of human biological differences. Just because two individuals might share similar dark complexions and have curly hair does not mean that they would be at equal risk of hypertension if one's genetic ancestry was adapted to the sodium-rich Atlantic coast of Africa and the other to the sodium-deficient interior of that continent.

My physician in Japan would not have been accustomed to working with the types of demographically diverse populations common to the practice of medicine in the United States. The Japanese, not unlike the Scandinavians whose staple is salted fish, had for centuries consumed 10,000 mg. of sodium a day. Therefore, it should not be surprising that medical laboratories in Japan assumed that my normal estimated glomerular filtration rate (eGFR) levels (measuring kidney function) were dangerously low. But as I would soon learn, American medicine had its own problems. The prevailing paradigm pathologized genotypical differences found in Blacks but not Whites and, as a consequence, missed, in some cases, the simplest and most obvious pathways to restored health. I wondered how many of America's 37 million Blacks of slave descent knew that they possessed one of the most precious gifts that nature could confer on living organisms—the attunement of their bodies over time to the ecological niche their ancestors had inhabited. There is no cure for kidney failure other than an organ transplant. But limiting how many pretzels and pieces of beef jerky to consume on a given day (once I became aware that, by over-indulging in such heavily salted foods, I am dishonoring the unique gift of sodium-retentive "super kidneys" inherited from my ancestors) is not a deprivation but rather a source of gratitude that brings me peace.

The more I dug into the medical literature, the more I became convinced that even more vital information was being missed because of America's two-hundred-year taboo on studying the history of the ancestral homeland from which our nation's entire population of enslaved persons/farmers had been taken. Ironically, an opposing taboo now hovered on the horizon, confusing the matter further: well-intentioned Americans who had taken a "let's be colorblind" approach to race in medicine opined that if race was a social construct and not biological, then studying biological and genetic traits might actually contribute to racial bigotry.

Even though our twenty-first-century medical establishment was now promising a new era of personalized medicine, something was wrong. For instance, the debate between colorblindness and racialized medicine was surely a dead end. One of the most unanticipated and perhaps politically unsettling discoveries of the Human Genome Project (HGP) was the fact that many, if not most, disease-triggering gene variants are population-specific. Ironically, this new finding has nothing to do with race. Human populations do exhibit differences, but in accordance with disease-causative gene variants that are localized and thus unable to match racial definitions (for example, West Africans from the non-cattle-breeding Tsetse Belt would not exhibit the same osteoporosis and other fragile bone disease risks as their East

African neighbors, who are dependent on pastoralism and a high-calcium diet). As a result of this confusion, African Americans of slave descent and other minorities were quietly being left behind since America's reference genome only identified those disease triggers most readily found in Europeans or other groups outside West Africa.

Undoubtedly, poverty and racial discrimination had a profound effect on racial health disparities. But the medical community was overlooking the fact that simple nutritional messages related to ancestral adaptations rather than normalized European standards could also save lives. Thus, it was the lack of ancestral knowledge regarding non-European Americans that could, at the very least, close some of the racial/ethnic disparities gap in the emerging field of genomic and precision medicine. While genetics is defined as the study of the role played by genes as units of heredity, genomics is the study of the composite profile created by an individual, community, or localized population's gene variants as they interact with one another and the environment.

THE APPROACH

Designations like race—a shorthand method of communicating the specific needs of particular demographic populations in our society—offer a useful organizing principle within American society. Thus, readers will easily recognize such common terms as Blacks, Whites, African Americans, Europeans, Asians, Latinos, Native Americans, or non-Europeans. However, and importantly, for any issues relating to specific genetic factors in medical research or clinical trials or genomics, I reject those terms. Here's why.

Twenty-first-century medicine requires precision, and racial terms are neither scientific nor precise. I recognize that many Americans have good intentions and believe we are really all the same under our different-colored skin. But I doubt whether those same Americans would show much patience if the laboratory values, clinical trial results, and pharmacological responses of White males were derived solely from medical research conducted among the Zulu of South Africa. It is this type of disparity that I aim to address.

In American society, those who narrate history are charged with telling stories while the medical profession is tasked with healing people. But sometimes new insights or solutions arise when people from different subject areas examine a problem. As a specialist in African historiography and evolutionary history, I see how past events lead to those in the present and how, if we don't correct our missteps, history can repeat itself. My training has also allowed me to develop an ability to look beyond specific details and see patterns—a skill I have used in the pages that follow.

In this book, I apply the tools of my discipline—posing different sets of questions and exposing different sets of etiological tracks—to help close the racial health disparities gap. These are not therapies, genetic analyses, or medical treatments of any kind. Rather, they are pieces of the story—evidence—that have been largely overlooked until now due to limitations in the methodologies of current medical research, which include:

- erroneous assumptions about race;
- lack of familiarity with novel ecological niches in Africa that demand unique genetic adaptations in certain populations;
- the tendency to pathologize certain neutral traits in localized populations solely because they have negative health consequences for Americans of European descent;
- the fact that certain areas of medical research in the United States are delimited by an unexamined but exclusionary

one-size-fits-all medical paradigm tailored to the biological needs of a single demographic population; and

- a failure to clarify the true nature of human genetic diversity to the general public.

Ecological Niche Populations (ENPs) was the name I gave to population configurations that shared a similar health disparity triggered by the same adaptive gene variant.

The following pages will explore the ways in which America's current approach to medicine employs a rigid paradigm that enlarges rather than closes the racial/ethnic health disparities gap. Our current medical system—and its genomic databases—does not yet have the data to adequately address certain diseases for which African Americans are at high risk. To be clear, my aim in writing this book is *not* to downplay some of the most exciting advances now being made in the genetic sciences. I am not a scientist. Rather, my intention is to apply my skills and experience as a historian of Africa and the African diaspora to an issue that affects me personally and that has touched—sometimes in heartbreaking ways—the lives of my family and many of my friends. My goal here is to contribute to the conversation on how genomic knowledge can be more accurately generated and applied to the health of Americans from *all* ancestral backgrounds. It is my hope that this effort will help other researchers, including the experts in more technical fields, recognize missed clues that may only be unearthed as we open ourselves to different ways of knowing.

OUR JOURNEY

We begin by looking at ancestral history and the ways in which certain gene variants found in African Americans of slave descent ensured their survival in the African interior but have since resulted in health problems in the United States.

In Chapter 1, we go back to a time when gold was sold by the sodium-deficient decentralized societies outside the boundaries of Timbuktu, Ghana, Mali, and Songhay for a commodity even more precious—rock salt from the Sahara. But only the elites could afford salt. Most individuals in those societies were subsistence farmers who never had the opportunity to take even a few licks of this precious mineral. We follow the history of these skilled agriculturalists who became vulnerable to slave raiders from the coastal West African states working in collusion with European and American slave-ship captains. We also explore the health consequences of a genetic population carrying the highly sodium-retentive African variants of the Apolipoprotein L1 gene, which allowed ancestral farmers to live on 200 mg. of sodium a day, but which became maladaptive once they were translocated to America.

Chapter 2 exposes the way flawed thinking about race, and new but counterintuitive discoveries made possible by the 2003 launch of the HGP, contributed to eugenics theories and held back medical breakthroughs for minorities in precision and genomic medicine.

Chapter 3 tracks the etiology of certain high-risk diseases in Blacks (such as hypertension, kidney disease, type 2 diabetes), high mortality in childbirth, and certain cancers (including triple negative breast cancer and metastatic prostate cancer) whose ecological clues have been overlooked because of American medicine's Eurocentric paradigm, which discounts genetic, ecological, and ancestral clues that are not directly relevant to improving the health of Whites.

Chapter 4 presents the taxonomic concept of the ENP model and its complementary algorithm. This model can be used to bypass racial designations in medical applications ranging from scoring organ transplant eligibility to determining those nutritional values in non-European Americans that differ sufficiently from the norm to trigger negative health consequences.

Chapter 5 details ten steps aimed at closing the disparities gap in medical care and advancing health for all. While most of the actions do not require large outlays of funding, they do demand a radical new outlook regarding the medical care of all populations. It is imperative that the medical community reposition itself to serve as the healthcare vanguard of a multigenomic nation.

1 OUR HISTORY

IN THE SHADOW OF TIMBUKTU

FLORENCE TATE—civil rights leader, Pan-African activist, writer, matron of salon culture, intimate friend, and surrogate mother/sister to me for forty years—died of kidney failure a week before Christmas in 2014. Even now, I struggle to find words to express the gravity of my loss. But the grief surrounding her medical condition—renal failure—was all too familiar; I was acquainted with far too many others who had suffered from it. One morning, several months after Florence's death, I sat at my kitchen table with a pad and pencil considering how many friends and family members had died of renal failure. Thirty minutes later, I had scribbled and jammed names into every available space on the page. The realization was alarming, but something more stopped me in my tracks. Although I have friends from many cultural backgrounds, the names on the piece of unlined paper, with few exceptions, were those of African Americans—most of whom had been under the age of sixty. While some went quickly, others were placed on dialysis machines. But all had succumbed to the disease.

Statistics from the Centers for Disease Control tell a more compelling story. Black Americans between the ages of thirty-five and forty-nine are 50 percent more likely than Whites to die from high blood pressure, 66 percent more likely to die from type 2 diabetes, and 100 percent more likely to die from stroke. By the age of fifty-five, 75.6 percent of Black American females suffer from hypertension (versus 40 percent of their White counterparts), a condition that leads to kidney failure at a rate three to four times that of other American demographic groups. And the mortality rate of Blacks from the types of hypertensions that trigger cardiovascular complications is several times that of Whites—one of the reasons for the unusually high death rate of Blacks during the early stages of the novel Coronavirus disease (COVID-19) crisis.[1] As a result, mortality rates of Blacks from COVID-19 before the advent of a vaccine were the highest of any ethnic group in the country. Importantly, many of the chronic illnesses that lead to higher mortality rates and have been linked specifically to COVID-19 mortalities—hypertension, strokes, kidney failure, and heart disease—are also directly linked to the metabolic processing of sodium in the body.[2] But why did so many in my community appear to have such a hypersensitivity to sodium?

THE LOST CITY OF GOLD

Throughout my undergraduate years, I spent one afternoon a week dragging around a backpack weighted down with the Arabic-English dictionaries of J. G. Hava and Hans Wehr. Walking from Harvard Square to an imposing brick building on Cambridge Street, I would arrive at the home of my Arabic professor, a German Jew who had narrowly escaped the Nazis. I had decided, aching from the loss of both of my Black grandmothers, that one could not have too many grandmothers. So Ilse Lichtenstadter became a loving *Savta* to me, and our relationship endured until her death in 1991.

One morning during my freshman year at Harvard, I mistakenly stumbled into Professor Lichtenstadter's Arabic class (thinking that I had signed up for a Swahili class). But Professor Lichtenstadter took me in and, from her lecturer's perch atop the Arabic section of the Near Eastern Languages and Literatures Department, guided my progress.

My passion for history in all its permutations emerged from a slow, labored translation of a book entitled *Al-Muqaddimah* (The Prologue) by the fourteenth-century Tunisian scholar Abd al-Rahman ibn Khaldun.[3] Ibn Khaldun, a philosopher of history, came to be known as the "Father of Historiography." He did not merely record historical events like Herodotus and Diodorus Siculus or chronicle the stories told to him about the past. Rather, he used history as an instrument of science to understand both the complexities of human nature and the ways in which its insights could be used for purposes of problem solving. With the patience of a lacemaker, Professor Lichtenstadter examined my work word for word and introduced me to the world of seventeenth-century Timbuktu and the empires of that region known as the Western Sudan. It was through these Arabic manuscripts that I learned details of the gold-salt trade.

The once-prosperous West African city of Timbuktu boasted one of the most active book trades in that region of the Islamic world. Founded in the twelfth century at the junction between the Sahara Desert and the valley of the Niger River, Timbuktu was tied into the vast Islamic trade network whose global reach included the importation of West African gold to Europe. At Timbuktu's height in the fifteenth and sixteenth centuries, wealthy families patronized Muslim scholars from West Africa, Morocco, Cairo, and other parts of the Islamic world. The texts being produced and copied ranged in subject matter from religious exegeses in Islamic gnosticism (that is, Sufism) to astronomy, mathematics, and historiography. In 1526, the Moroccan geographer Al-Hasan Muhammad al-Wazzan (known in Europe as Leo Africanus) published *The Description of Africa, Cosmographia et geographia de Affrica*, in which he reported, "We sell many [books] that come from the Berbers [Maghreb]. We receive more profit from these sales than from any other goods."[4]

Since the fall of the Songhay empire in 1591, Timbuktu has lain in ruin. Nevertheless, it was from these surviving manuscripts that I, as a graduate student in the early 1970s, came to understand the nature of the commerce that financed the immense wealth of the empires of Ghana, Mali, and Songhay. In recent years, one of the most promising developments that might serve to open this fount of knowledge to a larger audience was the 1997 delegation, led by Henry Louis Gates, Jr., of Harvard, to Timbuktu—which resulted in the establishment of one of Timbuktu's first modern libraries.[5] An estimated 300,000 manuscripts are believed to have been saved when in 2012 the city fell to Muslim fundamentalist militias, which were burning such books.[6]

As remarkable as this long and tormented history has been, my preoccupation over the past decade concerned the motivations behind this trade, which brought the empires of the West African interior into intimate commercial contact with the decentralized societies outside their borders. For the ecological details upon which this commerce was based offer previously hidden insights into the current state of African American health. However, the lost empires were not what grabbed my attention. Rather, it was the subsistence farmers who inhabited the decentralized but gold-rich regions outside the boundaries of these states. In the 1500s, Portuguese navigators were able to divert the gold trade from the Islamic empires in the interior to the Atlantic coast. And when the gold reserves finally petered out a century later, the subsistence farmers who once mined it had themselves become enslaved persons/farmers—better known in that region as "black gold." I am a descendant of this group.

These subsistence farmers occupied a unique ecological niche, a region so deficient in sodium that the elites were willing to trade their gold for rock salt imported from the halite mines of the Southern Sahara. The lower echelons of these societies perceived salt as a luxury item to be flaunted but not consumed. The communities without centralized governments were held together by common languages and shared cultural mores. But absent the military advantages of a state, the subsistence farmers were defenseless, given subsequent events.

The leading families in these communities, however, were able to pay ransom to slavers, either in cowry shells or in salt, to release their sons and daughters.[7]

Over the years, my research efforts periodically returned to this region, allowing me to pick up new tidbits along the way from the translated manuscripts over which I pored. The *Tarikh al Fattash* (The History of the Seeker of Knowledge) and *Tarikh al-Sudan* (History of the [Western] Sudan) both described the lucrative trade across the Sahara Desert, whose profitability had sustained ancient Ghana, Mali, Songhay, and Timbuktu. In perusing these pages, I became ever more intrigued by the history of Africa's interior.

Long before the slave trade, sea salt had been commercialized on the Atlantic coast of Africa using evaporation pans and boiled-water distillation techniques. But because of the tropical heat and humidity in the deep interior, this processed salt could not be transported without becoming contaminated by impurities from the sea water and thus unmarketable to the inhabitants. As a result, rock salt mined from deposits in the Southern Sahara became the principal commodity upon which the gold trade was built. But this commerce was controlled by elites in this region for whom the most precious marker of their status was the possession of rock salt crystals.

Even before Timbuktu rose to prominence, an Arab geographer from Cordoba, Spain—Ubaidalla Al-Bakri—offered a description of the earliest empire of this Western Sudanic region. In a book entitled *Kitāb al-Masālik wa-al-Mamālik* (The Book of Roads and Realms), he wrote, "Their king of ancient Ghana levies a transit duty of one gold dinar, a coin worth 4.25 grams of gold, on every donkey-load of rock salt brought into the country. He doubles the tax to two dinars when the salt leaves the country."[8] Five centuries after Al-Bakri's account, a scholar of Timbuktu named Abd Al-Sadi wrote in *Tarikh al-Sudan,* "Jenne on the Niger River near Timbuktu is one of the most important Muslim trading cities. It is where salt merchants from the mines

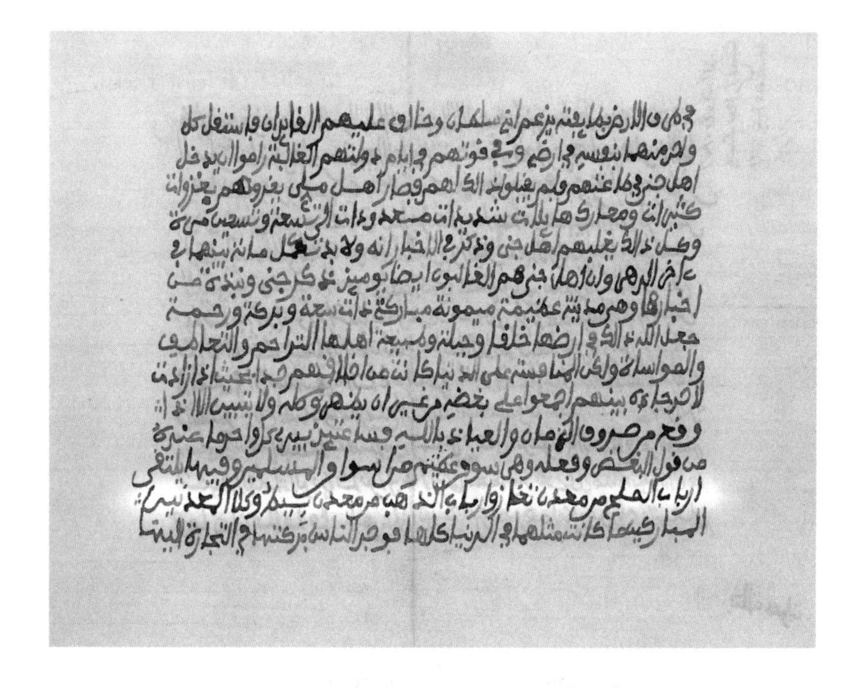

Figure 1.1 ***The Tārīkh al-Sūdān*** **(History of the Sudan) was written in Arabic around 1655 by the imam and chronicler of Timbuktu, 'Abd al-Rahman Ibn 'al-Allah al-Sa'di. This manuscript represents one of the most notable primary sources for the history of the Songhay empire.** Gallica, Bibliothèque Nationale de France, Paris (Manuscript B, Arabe 5256).

of Taghaza negotiate with gold traders from Bitu [on the northwest fringes of the Akan territories.]"[9] (See Fig. 1.1.)

Thus, Timbuktu was not the salt's final destination. Rather, it was transported to the Akan people who inhabited the forested region south of the ancient Ghanaian empire in a territory awash in goldfields. This trans-Saharan trade supplied nearly one-third of the gold used by European monarchs for coinage during the Middle Ages. Its far-reaching influence is evident in the Spanish language—the Spanish term for gold coin in the fifteenth century was *maravedi* (derived from Almoravid [*murabitun*] dinar).[10] Still, the Akan's profit margin during this early period in the gold trade was quite slim because of the multiple layers of middlemen that included local merchants, trans-Saharan caravan traders, and Mediterranean boatmen required to move the gold from the African interior to Western Europe (also referred to as Northern Europe).

Unlike the indigenous, hunter-gatherer populations of coastal America, South Africa, and Australia, the decentralized societies of the West African interior inhabited vast agricultural regions. The area was dotted with villages of subsistence farmers who developed deep spiritual ties to the land and, over time, complex agricultural techniques for turning hardscrabble, nitrogen-poor soil into crops sufficient to feed their families—even if the land produced no surpluses. The farmers worked hard simply to avoid starvation. Whereas in other parts of the world soil with low nitrogen concentrations was enhanced with manure as a natural fertilizing agent, the region could not support beasts of burden owing to illnesses caused by the tsetse fly and other tropical diseases, which made it impossible to breed farm animals. As a result, the agricultural societies were deprived of the fertilizing properties of manure and lacked the added muscle power of animals. (The lush imagery of African rainforests notwithstanding, this region still possesses some of the thinnest topsoil in the inhabited world.) Seasonal crops and rainforests require radically different environmental conditions to flourish. For these reasons, West African farmers honed a different set of skills to meet the challenges of agriculture in their ecologically unique and fragile environment. These same hard-earned skills would later make these enslaved persons/farmers exceptionally profitable workers to antebellum plantation owners.

As challenging as life was for subsistence farmers who inhabited regions that bordered the centralized states, they did enjoy one advantage over their surplus-producing neighbors: since there was little of worth to take, militaristic violence was minimal. That is, since these stateless societies could only provide sufficient food to feed each family in a village, their community was unappealing to empire-building states with invading armies. Land is valuable to human predators only when it is naturally capable of producing a surplus or contains other sought-after resources. This still holds true today, whether the resource is favorable soil conditions, precious minerals, or a particularly strategic location for purposes of trade.

Decentralized regions also had another protection—the use of "silent barter." This system of commodity exchange has been used over the centuries in other parts of the world where those engaged in swapping goods do not speak the same language. It developed as a cautionary measure when decentralized regions traded with potentially menacing centralized states such as ancient Ghana, Mali, Songhay, and Timbuktu. In the case of the African goldfields, Herodotus gave us a first glimpse of the barter system in Book IV of *The Persian Wars*, written in 430 BC. The Greek historiographer explains:

> Another story too is told by the Carchedonians [inhabitants of the ancient trading city of Carthage near what is today the city of Tunis]. There is a place in Libya, they say, where men dwell beyond the Pillars of Heracles [Greek mythology's term for areas beyond the limits of the known world]; to this they come and unload their cargo; then having laid it orderly along the beach they go aboard their ships and light a smoking fire. The people of the country see the smoke, and coming to the sea they lay down gold to pay for the cargo and withdraw away from the wares. Then the Carchedonians disembark and examine the gold; if it seems to them a fair price for their cargo, they take it and go their ways; but if not, they go aboard again and wait, and the people come back and add more gold till

> the shipmen are satisfied. Herein neither party (it is said) defrauds the other; the Carchedonians do not lay hands on the gold till it matches the value of their cargo, nor do the people touch the cargo till the shipmen have taken the gold.[11]

Farming communities uniquely suited to survive in an ecological niche that would not have been able to sustain most humans, including coastal Africans, evolved within West Africa's interior. That is, populations possessing the gene variants most critical to farming the hardscrabble soil in the tropical heat on a dietary intake of 200 mg. of sodium a day survived and grew. Lacking access to costly rock salt, the impoverished farmers of the West African interior flavored their food with the high-potassium mineral salts from burnt ashes of millet and other plants.

For centuries, Western European monarchs salivated over the prospect of one day seizing control of the massively lucrative African gold trade from the Muslims. Rumors filtered into Western Europe from the Arab world that the wealthiest man in the world was the Mansa Musa of Mali, a ruler dwelling somewhere in the interior of West Africa (Fig. 1.2).

Legend has it that Mansa Musa owned the largest stores of gold in the world. The lucrative trade routes that crisscrossed the Sahara Desert were well guarded by Europe's Arab competitors. It was the enterprising Portuguese navigators of the fifteenth century who so profoundly changed the fortunes of Western Europe for the better and the merchant empires of the Western Sudan (Ghana, Mali, Songhay) for the worse. Unable to compete with the Muslims for their overland trade routes, fifteenth-century Portuguese ship captains set out on the treacherous cross currents along the West African coast to search for the Mansa Musa's immense treasure in gold. As they became more adept as mariners, they hoped to prove the belief held by the first-century AD Greek geographer Strabo and Muslim geographers Ibn Khurdadhbih and Al-Mas'udi that Africa could be circumnavigated.

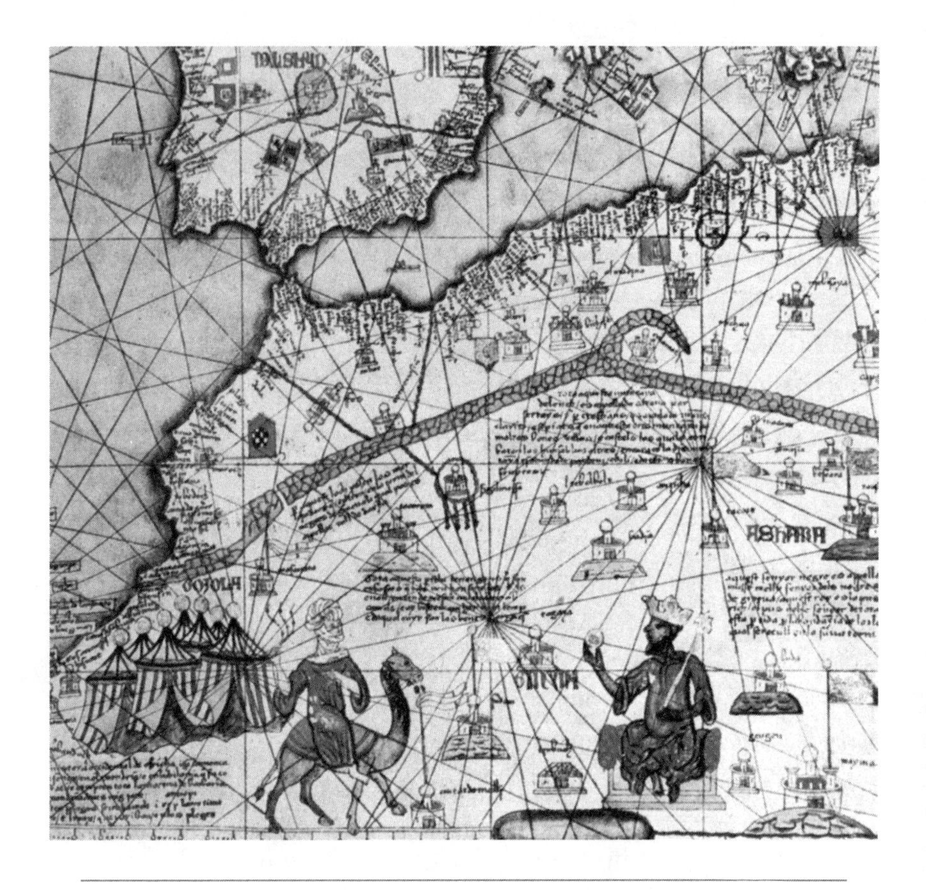

Figure 1.2 **This Catalan Atlas of 1375 depicts the "Golden Emperor" Mansa Musa of Mali.** CPA Media Pte Ltd./ Alamy Stock Photo.

By the seventeenth century, Western Europeans succeeded in wresting control of global commerce from the Muslims using the newly forged oceanic trade routes. What Europeans gained in this transaction would, in the end, flip their fortunes and place them ahead of their Islamic competitors. The search for maritime trade routes to the East led Western Europe to stumble upon and ultimately conquer the largest mass of fertile land in the world—the Americas.

Several decades before Christopher Columbus's 1492 trans-Atlantic voyage, the Portuguese mariners who had been plying the coast of West Africa in search of the Mansa Musa's fabled gold also found the treasure they sought. But just as Columbus had misnamed the inhabitants of the New World "Indians," the Portuguese initially assumed that the West Africans on the coast who appeared to have limitless access to gold were members of the entourage of the great Mansa Musa. The Portuguese traders had stumbled upon a centuries-old secret: the wealthy empires of the Western Sudan did not control the goldfields. Rather, as middlemen, they had built their empires on profits made from trading rock salt mined in the Southern Sahara with the decentralized Akan people. The Portuguese traders wasted little time establishing direct contact with the Akan, quickly diverting the gold trade from the Arab-controlled routes across the Sahara Desert to the Atlantic coast of West Africa. Within a century of this trade divergence, the last of the Western Sudanic states—Songhay—collapsed along with the trans-Saharan trade routes that had sustained the empires for centuries.

THE PAPAL BULL AND SLAVE PORTS

By 1537, Pope Paul III had become alarmed at the collapse of indigenous societies in the New World. The Spanish conquistadors invaded and conquered these regions, unleashing Afro-Eurasian diseases. The more isolated native peoples were genetically unprotected from these never-before-seen pathogens, including measles, smallpox, influenza, typhus, and tuberculosis. On June 2, 1537, the Pope issued a bull—a public decree—entitled the *Sublimus Deus*. It prohibited the enslavement of Native Americans, declaring them to be "rational beings with a soul." To respond to the ensuing labor shortage, Bishop Bartolomé de Las Casas, with the approval of Holy Roman Emperor Charles V, authorized the enslavement of sub-Saharan Africans.[12] The conquistadors considered the bull a veritable gift from God, as, after all, the West

Africans possessed the same level of innate immunity to agricultural diseases as Europeans and had adapted to tropical pathogens. They were highly skilled farmers whose ability to eke out a living in a nutrient-poor, low-sodium no-man's-land made them exceedingly marketable to plantation owners in the New World who had heretofore been unable to profit from vast tracts of fertile land because they lacked a skilled agricultural workforce to cultivate it.

Members of West African farming communities could be kidnapped with relative ease by slave traders because their societies, having been decentralized, were undefended. (It is understandable that modern observers of the trans-Atlantic slave trade's beginnings would view this tragic situation from a modern-day lens and assume that European racism against West Africans was the impetus for the brutal enslavement of the men and women loaded onto the slave ships, but the situation was more complex.) Contrary to the belief of many Americans, Black Africans were not as a general proposition kidnapped and shipped off to the Americas. The traders targeted skilled farmers from the decentralized states of the interior because they lacked governing structures that could protect them from foreign marauders. By contrast, inhabitants of the militarized states (also referred to as "Black Africans") on the West African coast had in place the same protective infrastructures to dissuade predators—armies as well as those seeking trade relations—as did the centralized governments in Europe and other parts of the world. For instance, in 1526, the king of the West African kingdom of Kongo—Nzinga Mbemba—who had adopted the Christian name Afonso I, upon learning that some of his citizens were being abducted by Portuguese traders, sent the following threatening letter to the Portuguese king:

> King João III, Sir, in our Kingdoms, I would like to call your attention to a matter that does not show honor and respect to God. There are many of my people who enjoy the wares

> and beverages of your Kingdom, which are brought to us by your traders. But in order to satisfy their greed, some of these merchants have been caught seizing some of our free people, who are neither captives nor prisoners of war. It has come to my attention that noblemen and sons of noblemen have been seized on our streets, kidnapped, and sold to the white men who are from your Kingdoms. The relatives of these kidnapped people have petitioned me with this problem, and thus they must be returned.[13]

The Portuguese king, not wanting to risk the lucrative trade relations that his country had with the Kongo, had no choice but to comply. But who would have the authority and economic leverage to return subsistence farmers seized from leaderless, decentralized societies in the West African interior?

The Akan states of the West African coast attained full military glory only when they were able to trade their gold directly to the Portuguese—because their profits increased ten- to twenty-fold. Traders from the Islamic empires of the Western Sudan had previously paid the Akan a pittance for this precious metal because profits were limited by the multiple layers of middlemen involved in the transactions. Initially, the Akan prospered. As we've seen in other regions throughout human history, the creation of surplus wealth drives the inevitable centralization of stateless societies into socially and politically hierarchical states defended by standing armies. However, the expansion of the gold trade depleted the ore in less than a century, and the newly enriched West African states on the Atlantic coast began desperately searching for a substitute product that the Europeans would value as highly as gold. Given the support of the Vatican and the immense demand for skilled farm labor in the New World, the trans-Atlantic slave trade exploded onto the landscape of West Africa with a new and even more profitable commodity—the aforementioned "black gold." Beginning in the mid-sixteenth century, coastal African slave traders swept into the decentralized societies of the West African interior and kidnapped and chained farmers and "delivered" them to European slavers on the coast—who stuffed the farmers into the holds of ships bound for the Americas (Map 1.1).

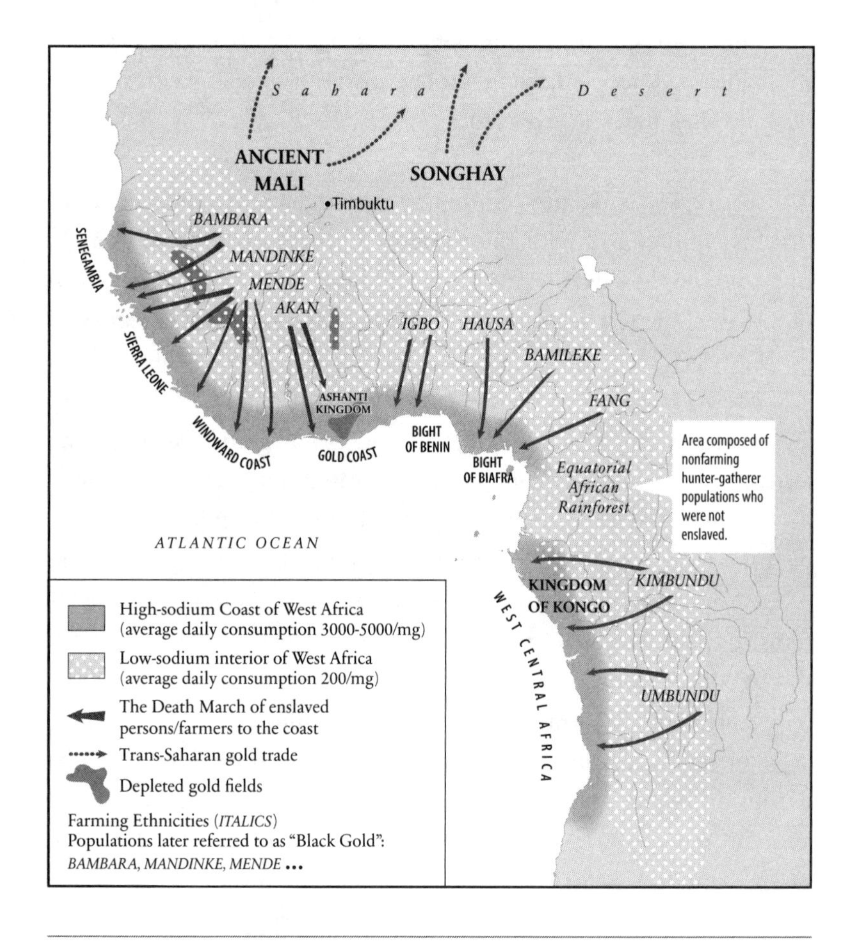

Map 1.1 This seventeenth-century map of West Africa records the 500- to 1,000-mile "death march" of enslaved persons/farmers from the interior to the coast, where they were loaded onto slave ships.

The European slave traders knew nothing of the origins or history of the enslaved persons/farmers they purchased at coastal ports with names like Ouidah or Whydah (also known as the "Door of No Return"), Lagos, Gorée Island, Aného (Little Popo), Grand-Popo, Badagry, and Luanda. Because the slave trade was both profitable and highly competitive, knowledge of the specific locales from which the men, women, and children had been snatched remained proprietary and was known only by African traffickers. (The only written documents of this time available to Western scholars were the slave-ship logs. However, these records noted the coastal port of embarkation for the Atlantic journey, implying that all enslaved persons/farmers were from the vicinity of the coast of West Africa.)

A few written records from European travelers did detail the scarcity of sodium in the West African interior. For example, the eighteenth-century Scottish explorer Mungo Park remarked that he "frequently and painfully experienced in the course of [his] journey" the lack of salt in the African interior.[14] Another European traveler to the West African interior, John Matthews, remarked unsympathetically about the slave trade:

> The best information I have been able to collect is, that great numbers are prisoners taken in war, and are brought down, fifty or a hundred together, by the black slave merchants; that many are sold for witchcraft, and other real, or imputed, crimes; and are purchased in the country with European goods and salt; which is an article so highly valued, and so eagerly sought after, by the natives, that they will part with their wives and children, and everything dear to them, to obtain it, when they have not slaves to dispose of; and it always makes a part of the merchandize for the purchase of slaves in the interior country.[15]

It is only now, in the era of twenty-first-century precision medicine, that the lack of clarity with which Americans had for centuries understood the details of the trans-Atlantic slave trade has proved costly. For nearly half a century, the medical community has devoted increasing amounts of research funding to a medical crisis among American Blacks: the unusually high rates of salt-sensitive hypertension and the mortality rate from kidney failure that is four times that of Whites. Yet there have been no breakthroughs. The lack of progress can be traced to one nearly ubiquitous stereotype about enslaved persons/farmers—that they came from the salt-rich coast of West Africa.

The lack of a full account of the trans-Atlantic slave trade has remained a gaping hole in the otherwise rich chronicles of American history. Without the knowledge of the ecological niche interior to which enslaved persons/farmers were genetically adapted and from which they were kidnapped and shipped to the coast, it has not been possible for medical researchers to perceive a critical truth—the populations that were shipped to the Americas as enslaved persons/farmers had emanated from a region to which they had, over the course of the millennia, become uniquely adapted. What has it meant for the medical community to lack the capacity to discern possible differences in disease risk between West African populations genetically adapted to diets of 200 mg. of sodium a day and their coastal neighbors, who, not unlike most Europeans, consumed up to 5,000 mg. of sodium a day?

HYPOTHESES

Salt was in general use throughout Europe for much of recorded history. It was either extracted from sea brine or the result of the mining of halite or rock salt. The salt-curing of beef and pork dates to ancient times. (Many English locales that end in "wich" [Middlewich, Nantwich, Leftwich, for example] are historically related to salt production.) It is therefore not surprising that the availability of this mineral on the European continent has, over time, led to most Americans adapting to a sodium intake of between 3,000 and 5,000 mg. a day.

By the 1980s, medical practitioners in the United States had begun to notice the unusually high rate of salt-sensitive hypertension in African Americans, a malady that often led to high rates of kidney failure and cardiovascular disease. Because American universities, acceding to pressure from the civil rights and Black cultural movements, only began to teach African and African American history in the late 1960s, knowledge of Africa, even among medical professionals, remains prosaic. The assumption that Blacks came from the Atlantic coast of West Africa negated any suggestion that their bodies would be genetically adapted to less salinity than those of Europeans.

In April 1986, Professor Thomas W. Wilson published an article suggesting that the "availability (of salt throughout the different regions of West Africa) was not uniform."[16] This statement was soon eclipsed by a hypothesis of Wilson's and Clarence Grim's—the Slave Ship Hypothesis—that was based on a particular type of selection bias.[17] Assuming that all humans possessed the same sodium-intake needs, it theorized, slaves who survived the trauma of the Middle Passage were those whose bodies proved to be more sodium-retentive. But absent clear evidence, this slavery-hypertension hypothesis lost credibility over time and, by the early years of the twenty-first century, had been categorized as urban myth.

In 2005, David Cutler, senior healthcare advisor to then-senator Barack Obama, economist Roland G. Fryer, and sociologist Nathan Glazer circulated a new theory positing that equatorial African populations have higher sodium-retention levels than other groups due to a sodium deficiency in their ancestral environment.[18] In competing with more exciting theories, though, this more accurate assessment of the genetic issues involved was soon set aside in favor of new theories linking hypertension in African Americans to either cultural factors, the stresses of systemic racism, or lack of preventative self-care. While, without a doubt, racial discrimination and other forms of institutionalized bigotry result in a range of stresses, such social determinants, however important, should not be deemed solely responsible for the hypertensive-triggered health disparities that dramatically increase chronic disease risks.

American medical textbooks mistakenly assert that 500 mg. of sodium a day is the lowest intake compatible with human life. The Yanomani Indians of the Amazon, for instance, inhabit an ecological niche bearing at least some similarities to that of interior West Africans. Unlike West African farmers, the Yanomani are an aboriginal population, but their ancestral diet included a daily sodium-intake level that was even lower than that of the ancestors of Black Americans—approximately 50 mg. per day. Today, like African Americans, the Yanomani suffer from unusually high rates of hypertension in the high-sodium food cultures in the urbanized areas to which they have relocated.[19] The Yanomani and interior West Africans share persistently high levels of circulating renin and aldosterone, which drive maximum salt retention in the kidneys. Referred to as the renin-angiotensin system (RAS) or the renin-angiotensin-aldosterone system (RAAS), these hormones regulate blood pressure, fluid, and electrolyte balance in the human body.[20]

Recently, geneticists have made progress in understanding the triggers for hypertension in certain populations. In particular, the proteins produced by the Apolipoprotein L1 (*APOL1*) gene have gained prominence as causative factors in Black Americans' high risk of end-stage renal disease (ESRD) as well as scarring of the kidneys known as focal segmental glomerulosclerosis (FSGS).[21] Given the paucity of genomic data relating to African Americans, certain gene variants may stand out in genomic studies only because they are rare in Europeans. That does not mean that they are triggering disease in Blacks. However, in the case of hypertension, researchers have found that the G1 and G2 variants of the *APOL1* gene are carried almost exclusively by populations of West African descent, suggesting that they are indeed responsible for the high rates of hypertension in Blacks. In 2019, medical researchers Mihail Zilbermint, Fady Hannah-Shmouni, and Constantine A. Stratakis published a review of sodium gain-of-function variants that were found in African Americans and West Africans but not Europeans.[22]

And yet, research conducted in the United States on racial and ethnic health disparities makes no mention of sodium as a possible causal factor. In fact, in the past forty years, hundreds of articles in medical journals have put forth a wide range of hypotheses to account for Blacks' unusually high rates of hypertension. The discovery that the West African variants of the *APOL1* gene increase sodium retention has thus far been shoved aside. Instead, the focus of *APOL1* gene research in the current medical literature has been on its ability to protect against trypanosomiasis (or African sleeping sickness)—a tropical disease borne by the tsetse fly that infects its victims with microbes of the species *Trypanosoma brucei* and, when left untreated, can lead to death. But while nobody in the United States dies of African sleeping sickness, an estimated 19,000 Blacks succumb to diseases related to high sodium consumption. Current medical research simply overlooks the unique ecological conditions of American Blacks' genetic ancestry. Instead, millions of dollars are allocated to studies that continue to attribute this population's salt-sensitive hypertension and kidney failure solely to stress resulting from racism.[23]

A MEDICAL MYTH?

The United States has the financial capacity to research hypotheses related to Black hypertension ad infinitum. But what it does *not* have is the motivation to communicate to the African American community that US Department of Agriculture nutritional guidelines that have been standardized based on populations of European ancestry may trigger sodium toxicity in those with a different genomic profile. And this situation does not appear to be improving.

A major meta-analysis published in 2011 and circulated in the national media claimed to show no convincing evidence that reducing one's sodium intake lowered the risk of heart attacks, strokes, or death in people with either normal or elevated blood pressure levels. The article, which called for further studies to clarify the new data, concluded that more randomized controlled trials would be needed to confirm whether restricting sodium may in fact be harmful to people suffering from heart failure.[24]

In 2017, James DiNicolantonio, a cardiovascular researcher and popular writer, asserted that a high sodium intake would both improve everyone's health and extend their life spans. The back cover of DiNicolantonio's book boasted, "In this book, a leading cardiovascular research scientist and doctor of pharmacy overturns conventional thinking about salt and explores the little-understood importance of it, the health dangers of having too little, and how salt can actually help you improve sports performance, crush sugar cravings, and stave off common chronic illnesses."[25]

Some nutritionists dismissed the dangers of sodium overconsumption and labeled it another medical myth.[26] New questions surrounding sodium intake arose in 2014 when several articles that derided the "unnecessary" stigmatizing of sodium in the diet began appearing in both medical journals and the national media. (In March 2014, an article entitled "Compared with Usual Sodium Intake, Low- and Excessive-Sodium Diets Are Associated with Increased Mortality: A Meta-Analysis" was published by a Scandinavian team.[27] Several months later, an article in the *New England Journal of Medicine* reported that a diet that included a sodium intake of between 3,000 and

6,000 mg. per day decreased the risk of death and cardiovascular events.[28] However, the same issue, devoted to the pros and cons of high sodium intake, contained several articles offering more conventional perspectives that linked high salt consumption to cardiovascular disease. Nevertheless, an NBC news headline that week announced: "Pour on the Salt? New Research Suggests More Is OK.[29])

With little pushback from the medical community, I decided to make my own voice heard. After several failed attempts, my brief letter to the editor was published in a 2014 edition of the *American Journal of Hypertension*:

> To the Editor: It is unfortunate that the Graudal et al. article "Compared with Usual Sodium Intake, Low- and Excessive-Sodium Diets Are Associated with Increased Mortality: A Meta-Analysis" does not contain a disclaimer (in bold type) stating that its findings should be disregarded by blacks. As researchers, we sometimes make the mistake of believing that inclusion in studies is the most meaningful way of addressing ethnic diversity in medical research. In some cases, such as in this study, it can be the best way to have a particular ethnic group's unique medical concerns erased. Because articles are appearing in the media based on the Graudal et al. findings asking whether low-sodium diets are just another "medical myth," black Americans may end up worse off than they otherwise would have been.[30]

I was grateful to the authors of the original article for offering a quick and thoughtful response, even if neither of our positions changed:

> RESPONSE: To the Editor: In the letter "Article on Sodium Intake Should Include Ethnic Disclaimer," Professor Hilliard states that "healthy blacks who consume more than 1,500mg

> of sodium per day are at risk for hypertension and chronic kidney disease" and that "accumulation of irrefutable medical data on the subject" shows that "American blacks are 4 times more likely than whites to die of kidney failure." The last statement may be correct, but the link to sodium is not established . . . *In conclusion, although Professor Hilliard's concerns are reasonable, the evidence relating sodium intake to health outcomes in blacks is sparse and equivocal and insufficient to support reliable conclusions* [emphasis mine]. We thank Professor Hilliard for providing the opportunity to emphasize that more studies in blacks are needed.[31]

Indeed.

African Americans are suffering from a health crisis for which too many vital clues are being overlooked. The reason is not that too little attention is being devoted to medical research aimed at closing the racial/ethnic health disparities gap. It is, rather, that the discoveries regarding human genetic diversity generated by the 2003 Human Genome Project are proving to be more unexpected and, in some cases, more unsettling than some Americans may be ready to accept.

2 THE SYSTEM

AMERICAN MEDICINE'S INCOMPLETE PARADIGM

THE HUMAN GENOME PROJECT (HGP), referred to in some circles as "the Book of Life," represented the largest collaborative biology venture ever undertaken. It deciphered the human genome by identifying the order, or "sequence," of its 3.2 billion base pairs, or letters. The HGP created maps identifying the locations of genes within major sections of our chromosomes and determined which genes on a particular chromosome were most likely to be inherited together. Based on the success of this international venture, it now seemed possible to apply biomedical data to the task of tailoring therapies in ways that met the individualized needs of patients. Thus it might be possible, for the first time, to identify the unique DNA sequences in individual cancer tumors as well as the mutations. In this way, the most effective pharmacological agent could be selected to block the cancer cells' mutagenesis.

Individuals in nonscientific fields, myself included, were ecstatic. The new ability of researchers to determine the sequence of nucleotide base pairs that make up human DNA would open the gates to the kinds of breakthroughs that would bring never-before-contemplated healing therapies to new generations. And if medicine was becoming individualized, we might be able to dispense with the volatile issues of race that had throughout our history blocked a smooth path toward civility in this country. Or at least that is what I thought at the time.

DEVALUING HUMAN GENETIC DIVERSITY

One of the earliest stumbling blocks emanating from the vast stores of new data generated by the HGP was both unanticipated and counter-intuitive. In fact, it made so little sense to the way Americans had always perceived race that, to this day, this disorienting concept remains a constraint on the growth of precision and genomic medicine in US minority communities. Fortunately, the problem does not lie in a lack of data. The evidence is unambiguous and uncontested. The hesitancy appears to come from the scientific world's reluctance to address the general public's deeply entrenched misinterpretations of human genetic diversity.

The headline of an October 2009 article in *Science* read "How *We* [emphasis mine] Lost Our Diversity: Human Ancestors Survived Two Genetic Bottlenecks as They Spread out of Africa." The article's awkward use of the royal "we" fails to include those of us whose ancestors remained in Africa (until, that is, they were kidnapped and involuntarily transported to the Americas). But it nonetheless offers what some nonscientists might consider to be a startling insight into the true nature of human genetic diversity, explaining:

> Researchers have known since the 1990s that Africans are the most genetically diverse people in the world. Humans outside of Africa are missing many genetic variants found only in Africans, and, indeed, the farther a traditional group lives from Africa, the less diversity it has in its genes and morphological traits, including skull shape. Genetic diversity is usually considered a good thing: the more a population has, the more likely individuals will be to have gene variants that will help them adapt better to new climates, diets, and life-threatening diseases, such as malaria or smallpox. Many scientists have suggested that those who left Africa went through a bottleneck,

where only a small number of individuals had offspring, thus reducing genetic diversity.[1]

Given the clarity with which the concept of genetic diversity was thus described, it was disappointing that over the course of the next decade the medical literature tended to diminish the role Africa played in our understanding of human genetics. Even prior to the launch of the HGP, a 2002 research team led by Ning Yu, a professor in the Department of Ecology and Evolution at the University of Chicago, had come to the same conclusion. In an article entitled "Larger Genetic Differences within Africans Than between Africans and Eurasians," Yu and colleagues described sequencing DNA segments in ten Africans, ten Europeans, and ten Asians. They found that the average nucleotide diversity was only 0.061 percent +/- 0.010 percent among Asians, 0.064 percent +/- 0.011 percent among Europeans, but almost twice as high—0.115 percent +/- 0.016 percent—among Africans. They concluded, "Africans differ from one another slightly more than from Eurasians, and the genetic diversity in Eurasians is largely a subset of that in Africans, supporting the out of Africa model of human evolution."[2]

Recent studies now concur that the Khoisan of South Africa carry the largest genetic diversity of any human population. What this compilation of data means in simple terms is that the African genome encompasses the full range of human genetic diversity. But how is that possible? When SLC24A5, or the so-called European white-skin gene variant, was found in the Khoisan ethnicity, the first theories suggested that this Southern African ethnicity had only appeared in this African genotype after the seventeenth century because of intermarriage with Dutch settlers on the Cape of Good Hope in South Africa (who later became known as Afrikaners if they could "pass" for European or mixed-race or "Coloureds" if they could not). When further investigation showed that the SLC24A5 gene variant was at least 2,000 to 5,000 years old, new studies appeared tracing its origins in Southern African populations to Middle Eastern migrants, even though the evidence of such gene variant transfers remained speculative.[3] The HERC2 and OCA2 genes, associated with light eyes and skin, have also been found among the Khoisan. Even though this population carries humanity's oldest genetic lineages, some members of the scientific community continue to maintain that most European genotypes differ from those of Africans rather than constituting a subset of the parental genome. Such attitudes might also explain the lack of motivation among Western scientists to sequence more than a fractional 3 percent of the African genome. (See Figure 2.1.)

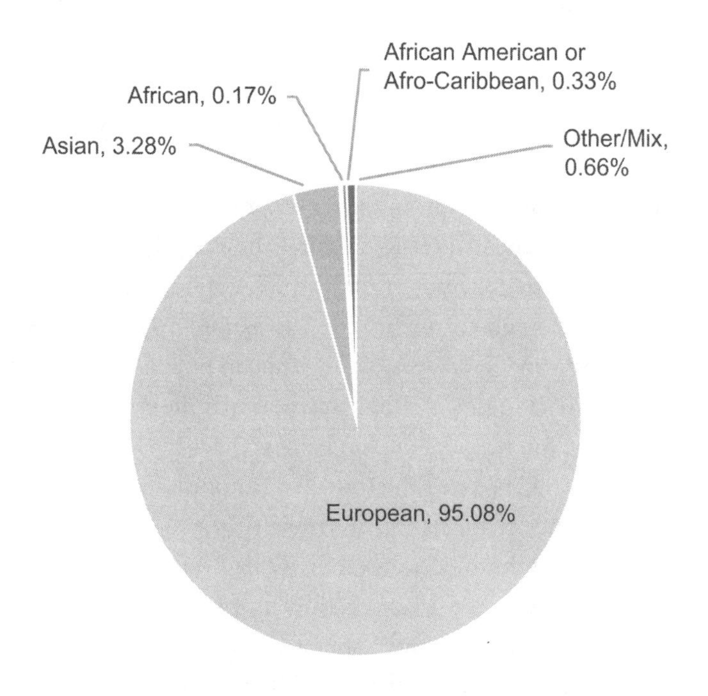

Figure 2.1 Diversity of participants in Genome-Wide Association Studies. *Data source:* GWAS Diversity Monitor, https://gwasdiversitymonitor.com, June 2023.

Even at the insufficient level at which the African genome has been sequenced, the genetics community has reached a consensus that it holds the greatest range of human genetic variation. As long as scientific funding agencies in the West maintain the knowledge base regarding the variants contained in the parental African genome at such a minimal level, eugenics-minded researchers can continue to insist that pale skin and other gene variants associated with white-race phenotypes originated with migrants to Europe and the Middle East. The more compelling evidence that these variants are also found among the oldest branch of humans is dismissed as mere speculation. Essentialist theories that give centrality to the European genome will remain in textbooks until more committed efforts are made to study and sequence the entirety of humanity's African parental genome.

In any case, this is how racial essentialism works. An elaborate taxonomy of skull shapes developed in the nineteenth and early twentieth century asserting that the Aryan white race possessed a dolichocephalic (long and thin) skull shape while lesser-developed humans had a brachycephalic (short and broad) skull shape.[4] We now know that the African continent, because of its greater genetic diversity, boasts skulls of all shapes and sizes. A similar correction has occurred over the past decade regarding the inference that Europeans were genetically a step above Africans because the former had Neanderthal DNA and the latter did not. In fact, the paleoanthropologist John Hawks from the University of Wisconsin called the notion that the presence of Neanderthal DNA correlates to elevation of a race "mythology" and, in a 2021 article, asserted, "At the time that a draft Neanderthal genome was published, a myth became established among the public that today's Africans are different from all other living humans in that they lack Neanderthal ancestors." Hawks added, "This notion is incorrect . . . The amount of Neanderthal genetic ancestry in African populations is basically what would be predicted from the extent of haplotype [an inherited cluster of gene variants] sharing between Africa, Eurasia, and Oceania, reflecting historic and prehistoric gene flow between these regions during the last 50,000 years."[5]

Interestingly, the latest research suggests that the Neanderthals and Denisovans, once believed to have evolved outside of Africa, have an African past and may have shared their light-skin gene variants with the Khoisan. In 2022, an article entitled "How Neanderthals Became White: The Introgression of Race into Contemporary Human Evolutionary Genetics" by Portland State University Gender Studies Professor Lisa Weasel was published in the *American Naturalist*; it highlighted the explosion of research on the Neanderthal genome after it was first sequenced in 2006 and before researchers learned that this DNA was not exclusively European. Weasel noted, "Most of these studies have focused on pale skin as offering a selective advantage to the levels of ultraviolet light radiation and associated vitamin D synthesis experienced by humans living at higher latitudes, specifically in Europe following migration out of Africa."[6]

It is reasonable to assume that de novo mutations created "the White race" after this branch of humans migrated out of Africa and settled in northern climes. However, the genetic evidence now shows that many "racial phenotypes" are in fact variants embedded in the African genome, which is parental to us all. These variants lie dormant, unexpressed, until a population's relocation to a challenging new environment requires expression of such traits as lighter skin color and straighter hair—which offered survival benefits such as greater penetration of the sun's ultraviolet light into the body. But this understanding represents a scientific consensus seldom conveyed as such to the science-reading public. For instance, the headline of a March 2022 article in *Science* read, "The Complete Sequence of a Human Genome." However, buried in the article's text and graphs was the boilerplate language that I had over the course of a decade come to dread: "Although CHM13 represents a complete human haplotype, *it does not capture the full diversity of human genetic variation. To address this bias, the Human Pangenome Reference Consortium has joined with the T2T Consortium to build a collection of high-quality reference haplotypes from a diverse set of samples*" (emphasis mine).[7]

Two decades after the launch of the HGP, the field of genomics has sequenced a mere 3 percent of the African genome—with the exception of those gene variants also found in Europeans. Efforts to "correct" the reference genome have induced Genome-Wide Association Studies (GWAS) to update the material nineteen times by 2019, with the latest update referred to as GRCh38. But most GWAS efforts focused on closing many of the 150,000 gaps in the original reference genome rather than further diversifying it. Big Data processes are typically employed for endeavors that promise significant market value. Little pressure was felt to evaluate achievements on a broader public stage. After nearly two decades of advances in the field of genetics, no authority was willing to admit publicly that the original dream of a "universal" genome had been overly ambitious. Instead, we find a growing number of research projects purporting universality with the use of such terms as "panhuman" and "pangenome." But the fact that only 2 percent of the African genome had been sequenced by 2011 and the percentage had only risen by an additional 1 percentage point in 2018 should alert us to the ongoing oversight.[8]

Given that genomic science is based in the Western world, it is not altogether surprising that the African genome has been severely understudied. But one of the most meaningful discoveries in the field is the centrality that the African genome will play in ensuring human survival from future lethal pathogens. That is, the only chance our species might have for surviving a devastating and rapidly spreading pandemic that kills Eurasians, Latinos, and indigenous populations will be whatever natural immunities are embedded in the more-extensive gene variants found on the African genome. However, in the West, the continent of

Africa remains marginalized because of its brutal history of colonialism, widespread poverty, and political instability. The genomes of its inhabitants are assumed to be replete with inconsequential and even primitive gene variants. Given the "otherness" associated with African ancestry, it is not surprising that the scientific community reflects the biases of the larger society. But the truth is that the African genome, far from being marginal, is the closest thing we have to a genuine panhuman genome (as it represents the entire ancestral DNA lineage of our species). And let's under no circumstances confuse "ancestral" with "archaic." These gene variants are alive and transferable with the same ease and vigor as all human mating behavior and contain natural immunities to ancient lethal pathogens that could recur at any time.

The true nature of human genetic diversity can be most easily understood by making geographical comparisons. The East African countries of Kenya and Tanzania are neighbors. If racial essentialism (that is, the belief that racial groups form discrete genetic divisions) was an accurate descriptor of human variation, a Kenyan and a Tanzanian, both racially categorized as "Black," would have more gene variants in common with each other than either would with a random European. But such has not turned out to be the case. The same gene variants carried by Europeans are scattered among inhabitants of the African continent. But the gene variants carried by either the Kenyan or the Tanzanian that are not also found in the European will probably not be found in Europe—or, for that matter, in Asia, Oceania, or among indigenous Americans. The further an ancestral population has migrated from Africa, passing through ever more bottlenecks, the smaller its range of genetic variants. And yet, the frameworks used in science play such a defining role in research that they silence any challenge to the given rules of a particular model.

So how is this admittedly confounding diversity principle treated in the scientific literature? The authors of an article entitled "Human

Pangenome Project: A Global Resource to Map Genomic Diversity" explained that "a 'pangenome' is the collective whole-genome sequences of multiple individuals representing the combined genetic diversity of the species." However, they then state, "The initial HPRC [Human Pangenome Reference Consortium] dataset cannot be comprehensive of global genomic variation, but it can set a foundation to build on. The HPRC will initially produce high-quality genome data for 350 individuals (700 haploid genomes) selected to maximize global representation within the logistical constraints of the initial HPRC efforts."[9]

The largest degree of human genetic variation will not be found by combing the globe for understudied tribes and ethnicities. The last migratory branches of our species will have the least amount of genetic variation because they will have passed through more bottlenecks, and the newly branching community will retain fewer variants than their familial branch. If the structure of human genetic diversity is not racial clusters, what is it?

NESTED SUBSETS

In the history of modern science, few visual models have offered the conceptual clarity required to make sense of complex molecular behavior—with the exception of the "double-helix" model. This facsimile of DNA structure, widely considered one of the most valuable scientific discoveries of the twentieth century, earned molecular biologists James Watson and Francis Crick the Nobel Prize in 1962. In later years, Watson lost several honorary titles and some degree of prestige for making controversial statements on subjects ranging from race and sexuality to gender and religion. However, those statements could not diminish the fact that the double-helix model had advanced the scientific community's ability to visualize new and innovative approaches to the field of genetics.

Perhaps with less fanfare than the discovery of the double-helix model will be the contribution that the concept of "nested subsets" makes to Americans' understanding of race and human genetic diversity. And yet, despite conclusive evidence to the contrary, the traditional, "essentialist" concept of race continues to dominate public discourse. (See Figure 2.2.)

"Essentialism" claims that there is a biological (or genetic) essence that defines all members of a racial group. Thus people in Europe have pale skin and lighter hair because these mutations helped ensure survival outside Africa and quickly became dominant. This theory of genetic diversity makes perfectly good sense and confirms our visual perceptions of reality. Academics and the general public have for years increasingly woven together assumptions based on the essentialist premise. However, the possibility that this belief system did not accurately describe the workings of nature first entered into the debate nearly two decades ago.

In November 2003, just months after completion of the HGP, Jeffrey Long, chair of the Anthropology Department at the University of

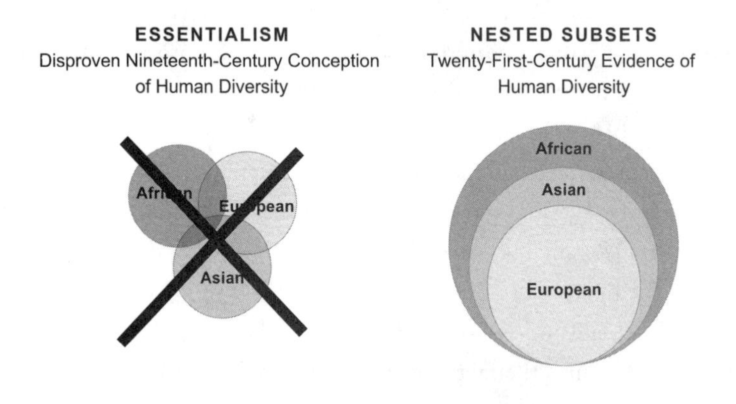

Figure 2.2 **Beliefs in "racial essentialism" form the basis of eugenics and other theories of racial superiority.**

New Mexico, raised an alarm at the annual meeting of the American Anthropological Association. He explained that the advances made by the HGP were just a start because they did not take into account the fact that African populations carried possibly twice as many Single Nucleotide Polymorphisms (SNPs) gene variants as non-African populations. If the standard genome developed by the HGP was based on European DNA, it would at best carry only half the totality of human gene variants. According to Long, "Most nucleotide diversity in non-Africans is a subset of nucleotide diversity in Africans. An interesting implication of this finding is that most common variation in the genome could be identified by studying a sample composed only of Africans, but a good deal of the common variation would be missed by studying a sample composed only of Europeans or Asians."[10]

Geneticists Jada Benn Torres and Rick Kittles further clarified the nested subset concept in a 2007 paper in which they observed, "Most of the alleles that are common in non-African populations are also common in African populations. Thus, the pattern of genetic variation is one of nested subsets, such that the variation in non-African populations is a subset of the variation found in African populations. The out-of-Africa model, which postulates an African origin of modern humans and subsequent migrations out into Eurasia and the Americas, best fits with this genetic data."[11] While scientists continue to debate the concept of nested subsets, an even more puzzling conundrum relating to human genetic diversity may finally be beginning to make sense.

THE LEWONTIN PARADOX

One wintry fall morning in 2011, I hurried up Oxford Street in Cambridge, Massachusetts, toward a Georgian brick building with the words "Museum of Comparative Zoology" carved into the lintel above the entranceway. As always, I felt a fleeting sense of profound unease when I entered this museum once named for Louis Agassiz—the nineteenth-century Swiss-American zoologist who achieved lasting fame for discovering the Earth's last Ice Age, classifying extinct species of fish, and theorizing that Blacks were a separate, inferior species relative to Whites. But my errand that day was to drop off the final chapters of a manuscript at the office of retired Harvard geneticist Richard Lewontin, who had generously offered to read through my work. The manuscript, entitled "Straightening the Bell Curve: How Stereotypes about Black Masculinity Drive Research on Race and Intelligence," explored the sometimes-veiled syllogism of "brawn versus brains" that drove a group of otherwise-rational male scientists to embark on eugenics research. Lewontin was revered in the scientific community as one of the most innovative geneticists of his time. In 1972, his landmark paper, "The Apportionment of Human Diversity," found more variation within so-called racial groups than between them, leading him to argue that such distinctions had no genetic or taxonomic value.[12] His argument, penetrating and new, used scientific data to confirm that traditional racial groupings seemed incapable of accounting for the amount of genetic diversity observed in the human genome. Two years later his book *The Genetic Basis of Evolutionary Change,* laid the foundations for the dynamic new field of population genetics.[13]

Lewontin's pithy but gently stated advice to me that fall morning—"Just remember, Miss Hilliard, the races are not real. Think of them as optical illusions, albeit wily, tricky ones"—nudged a latent gear in my psyche. Over the years, I had paid little attention to references made in scientific journals regarding a technicality called "Lewontin's Paradox." But by 2022, even though the Harvard geneticist had passed away the previous year, a scientific stew was in full boil. I took note of the fact that, in 2022 alone, Google Scholar cited 367 articles in its database using the keywords "Lewontin" and "paradox."

The uproar centered around Lewontin's claim that there was no correlation between the amount of genetic diversity within a species and population size. The observation appeared not to make logical sense—because genetic diversity (or variation) represents the axiom upon which biological life is based. The greater the number of gene variants in a genome, the better able that species will be to survive and adapt to changing environments, whether ecological, pathogenic, or climactic.[14] The result would logically be population growth. It was only in working on this book that the relevance of Lewontin's Paradox came into clear focus for me. Maybe it is only an enigma to us that nature would favor genetic diversity over population size because we as humans have different standards for evaluating what's important. The narrative of human history remains that of exalting the predatory power of empires over the defenselessness of sparsely populated decentralized societies.

There are only half a million Khoisan people in the world. But members of their community carry the most comprehensive range of gene variants. The remainder of the seven billion humans on this planet share half as many of these variants—which is, in essence, Lewontin's Paradox. That is, an African community that consists of only 0.00007 percent of the entire global human population carries more genetic diversity than the other 6,999,500,000 humans. If a lethal new pathogen against which some humans carried a form of natural immunity arose, the Khoisan would have a 100 percent greater likelihood of surviving the effects of such a pathogen than the rest of us. In fact, if we look closely at epidemiological history (or even just COVID-19), a rather shocking imbalance emerges. It might take days or weeks for a lethal pathogen's mutations to batter down our natural defenses or even those of laboratory-manufactured vaccines. But on the other hand, in 500 years of exposure to measles and chicken pox, indigenous populations have still not developed natural immunities to these diseases.

When scientists speak of genetic diversity creating population resilience, they are referring to genetic adaptability to environmental and climactic factors as well as lethal microbes. However, any number of catastrophic events, from earthquakes and tsunamis to an enemy's precision bombs, will result in the survival of a larger population, regardless of the number of gene variants that population carries.

It is the issue of genetic diversity (that is, the level of genetic variation in humans) that causes the greatest degree of confusion regarding popular perceptions of race. This is because certain gene variants carried by an individual may or may not be visually expressed. When individuals are racially categorized based on outward physical traits, the assumption is that people displaying different traits carry different sets of gene variants. However, our evolutionary history shows that the actual story of human genetic variation is not a visual one. Its plotline cannot be surmised by the visual traits we note in one another. The branching-off processes by which humans came to inhabit much of the globe can be tracked through genetic analysis, not by the way we look. Why? When a segment of early humans migrated out of Africa more than 60,000 years ago, a bottleneck was created. That is, whatever percentage of the original community the migrating branch composed, it would only have retained a percentage of the 324 million gene variants carried by the parental community. (This is not to be confused with the fact that all humans carry the same number of genes–approximately 20,000.) But branching populations will not carry the full load of random variants found in the original community. The concept that the new community will be genomically defined by the smaller migrating population's variants is known as the founder's effect. Population size, on the other hand, is not to be confused with genetic variation or the number of gene variants carried by a population. The expansion of a population into a new ecological niche is a function of the resources available to sustain larger numbers of people. But it substantially affects the dimensions of the gene pool only if outsiders enter the community.

Given the popularity of ancestry testing firms, it is not surprising that the general public has been led to believe the falsehood that people can now be racially categorized with even greater precision than would have been possible decades ago. We must remember that gene variants related to so-called Eurasian phenotypes such as pale skin, light hair color, and eye shape and color are nestled within our African parental genome. The biomarkers used by DNA testing companies to differentiate ethnic and geographic populations in different countries are not those related to skin, hair, and eye color because such phenotypes are far too common and encompass billions of humans. Rather, these companies find rare variants that have no obvious visual manifestation but that can, for instance, determine that an individual of Mexican descent's biogeographical ancestry is Zacatecas mestizo or Zapotecas Amerindian and that an African American may have had ancestors from Hausa, Ibo, and Malinke in West Africa and has Scottish, English, and Welsh DNA.

WHO WILL SURVIVE CLIMATE CHANGE?

In 2021, Minke Holwerda, a medical student in the Netherlands, noted a recurring theme in headlines such as these around the world: "Russia Anthrax Outbreak Affects Dozens in North Siberia" (BBC News); "The Permafrost Pandemic: Could the Melting Arctic Release a Deadly Disease?" (Greenpeace); "Deep Frozen Arctic Microbes Are Waking Up" (*Scientific American*); "All Hell Breaks Loose as the Tundra Thaws" (*The Guardian*); "How Thawing Permafrost Could Resurrect Long-Dormant Diseases" (*Arctic Today*). His ensuing article, "Pathogens in Permafrost: The Next Pandemic?" presented a nightmare scenario that focused less on naturally thawing viruses than on overly ambitious medical researchers who overlooked the potential infectivity of the viruses.[15]

Holwerda's work propelled me to take more seriously the threat to humankind implied in the Lewontin Paradox. The question that the late geneticist posed was simply, how is it possible that the number of gene variants in a genome will play a more significant role in human survival than population size? The only weapon that future humans might have against the release of ancient pathogens, such as those that might be unleashed in the melting permafrost, or even new ones, will be found, if anywhere, in our human DNA encyclopedia—the African genome. But why must we look to Africa? Most natural immunities will have entered our species during a similar microbial crisis possibly even millions of years before we turned the evolutionary corner and became homo sapiens. Whichever human community has retained the largest number of our species' variants may have the best chance of bringing to the fore, and in a sufficiently short period of time, a recessive gene variant that could better adapt to a new pathogen or other radical environmental change. This would be the case despite the enormity of the population on the other side of the equation, which would possess fewer gene variants available to confront nature's unknown challenges.

THE STRATIFICATION DILEMMA

Although more attention needs to be given to the African genome, we must nevertheless bifurcate our research agenda. The importance of the African genome is theoretical. It will advance genetics in ways that would appear magical given our biological knowledge today. But that goal should not be confused with the more immediate clinical importance of aggregating localized data on disease-triggering variants in discrete and relatively small demographic groupings. That is, theoretical research into the African genome will not, in the short run, cure shingles or irritable bowel syndrome. Advances in clinical research, on the other hand, will require the creation of databases containing precise population-specific data with stratification rather than diversity in mind. But progress in this arena has thus far been more rhetorical than real.

In the latter part of the twentieth century, the theories of scientists like William Shockley and Arthur Jensen, who made essentialist arguments claiming to prove the biological distinctiveness of the Negroid, Caucasoid, and Mongoloid races, rose to the fore. African Americans' susceptibility to sickle cell anemia was often referred to as proof that inherited differences between Blacks and Whites could be translated into a cognitive hierarchy with Whites at the top. Those scientists were unaware that sickle cell anemia is a byproduct of natural immunities to malaria and is therefore found in Southern European countries and parts of India. It should therefore come as no surprise that the disease is not present in nonmalarial regions of Africa.[16] The legacy of racist science in the United States left an indelible stamp on the consciousness of contemporary Americans. Well-intentioned efforts to heal the wounds of the past created a new set of barriers to closing the health disparities gap between the American majority and minority communities. Such efforts created an emphasis on "colorblindness," which downplayed the value of discerning localized demographic patterns in disease susceptibility, which had little if anything to do with broadly defined racial categories.

Sickle cell anemia is an autosomal recessive disease (that is, risks are inherited and, therefore, differ according to designated populations). It affects one of every 365 children born to American Blacks. But even with that being the case, the taxonomy appears to be racial only because Africans outside of Africa's malarial zones were not kidnapped and shipped to America during the trans-Atlantic slave trade.

Whites, by contrast, carry a 1 in 25 risk of inheriting cystic fibrosis and a 1 in 47 risk of inheriting spinal muscular atrophy (SMA). (Cystic fibrosis and SMA are disorders that affect the motor neurons of the spinal cord and brain stem.) Tay-Sachs is an inherited, neurodegenerative disease carried by 1 in 30 Ashkenazi Jews.

Today, the medical community does not contend that testing Blacks for the sickle-cell trait, Whites for cystic fibrosis and SMA, or Ashkenazi Jews for Tay-Sachs should be condemned and outlawed as "racialized medicine." Given the demographics of American society, to test all newborns for all four of these diseases would be an inexcusable waste of healthcare resources. The real problem is that the popularly used racial delineations lack sufficient precision to serve as the basis of medical diagnoses and can, at times, be dangerously misleading. Twenty-first-century genomics has given the medical industry an opportunity to jettison race from its arenas of research because we now have the precision tools to test DNA ancestry. While it is true that racial terminology and ancestry can overlap, when the stakes are at their highest, they sometimes don't.

Four years after the launch of the HGP, the first large-scale genome-wide association study identifying potential disease-causative gene variants was published.[17] This analytical tool is used in genetic and genomics research to associate specific genetic variations with particular diseases. The method involves scanning the genomes from a cohort of people and looking for genetic markers that can be used to predict the presence of a disease. Once such genetic markers are identified, researchers can seek pharmacological agents that can block the action of the destructive variants.

At first, the GWAS succeeded, beyond the wildest dreams of its creators, in scanning the genomes of large numbers of individuals and finding genetic markers or variants that could both predict the presence of medical conditions and develop treatment strategies.

Immunotherapy and precision oncology embraced this methodology and emerged as the most financially successful branches of genomic medicine, focusing primarily on sequencing cancer-tumor DNA in search of individualized cures. Such methods far outstripped reliance on one-size-fits-all chemotherapy treatments and generated hundreds of billions of dollars in profit less than a decade after release of the completed reference genome. Pharmaceutical companies that manufactured precision drugs aimed at fighting tumor cells advertised that they could tailor these medications to an individual's unique cancer profile. Seemingly overnight, this area of cancer research became a commercial powerhouse. But the market for therapeutic products turned out to be embarrassingly exclusive at a time in our history when many Americans harbored hopes that our nation might finally have become post-racial.

The efforts of the GWAS to match disease-causative variants in populations of individuals of African descent were a failure, with genetic sequences of these minorities' tissue samples coming back from the lab tagged as "Variants of Unknown Significance" or "Rare." This was about race in the sociological sense of the word because those who were failing to benefit from the innovations in precision medicine were Americans of slave descent. But the issue could at the same time be seen as a repudiation of race because there was no way to know if the failed genetic tests involved any group of Africans other than those from a well-defined region within the continent's interior. Science reporter Brian Resnick, in a *Vox* article, noted a disturbing fact that had been glossed over for years by the scientific community, even in discussions about improving diversity: "For these studies to work—that is, for them to yield the genetic markers that most accurately predict biological outcomes—researchers have to limit subject pools to people of a common ancestry. This allows them to control for the SNPs that have nothing to do with the disease or trait they're studying."[18]

An analysis conducted in 2009 revealed that 96 percent of participants in GWAS were of European ancestry. By 2016, slow and steady progress had reduced that number to 80 percent, but it did not reflect much progress for Americans of African descent. Instead, genomics researchers were taking advantage of the fact that the governments of China and other Asian countries had expanded genetic analyses of their populations and thus added their non-European genomes to the GWAS database. Not only were populations of African descent being left behind, but desperately ill Black Americans were being encouraged to take the same costly lab tests as everyone else, when the results might only benefit them in proportion to the European DNA they carried in their admixed genome.

The continuing search for human gene variants had exploded the number found from 2 million in 2001 to 324 million by 2019, but this only complicated the diversity problem.[19] Individuals of African descent carried the largest number of these gene variants (sometimes referred to as alleles or Single Nucleotide Polymorphisms), but only a minuscule number of their genome had been sequenced or studied. Far more staggering was the discovery that the theory that dominated medical genomics during its early years—the Common Disease, Common Variant (CDCV) hypothesis—turned out to be deeply flawed. The hypothesis postulated that the same disease-causing variants would be found in all human populations that manifest certain disease symptoms.[20] The sluggishness with which the medical community is awakening to this erroneous presumption is proving catastrophic to efforts to close the health disparities gap that continues to disadvantage minorities. Until now, gene variant matches that may have cured certain cancer tumors or other diseases only proved effective on Whites and ethnicities with high ratios of admixture with that genetic population's DNA. Also, while the distribution of disease-causative variants is a genomic process, it could not be identified racially or even by dividing populations according to continental ancestry.

Groups with the same phenotypes or physical traits did not necessarily share the same disease-triggering variants.

Researchers from China may have been the first to acknowledge this reality by regrouping populations according to shared disease variants rather than shared ethnicity. For instance, they developed distinct reference genomes for the Northern Han, the Southern Han, and the Beijing Han. The reference genomes corresponded to the different sets of gene variants that appeared to trigger the same diseases in each regional population. But researchers in America were so deeply committed to diversity that they proceeded to add the DNA of minorities to the predominantly White reference genome. Entangling so many different gene variants from diverse populations suffering from the same diseases tended to advantage the reference genome with the greatest degree of homogeneity—that of Whites.[21]

Geneticist Sarah Tishkoff of the University of Pennsylvania Medical School pioneered much of the groundbreaking work in repudiating the CDCV hypothesis in favor of tackling the far more complex matter of identifying population-specific gene variants. Tishkoff explained, "[My] findings further undermine the idea that common diseases are caused by common variations. . . . When genomics researchers first looked at the genome for links to diseases, that was their assumption, but they quickly saw it fall short."[22]

Conducting field work in Africa in 2008, Tishkoff began by tracking the etiology of lactose tolerance. Her research identified the gene variant responsible for lactose tolerance in Northern Europeans as being the lactase-phlorizin hydrolase (*LCT*) gene 3910-T/T gene variants. Among Eastern Europeans, the Kazakhstanis, and populations inhabiting Northern India, the same function is performed by the *LCT* gene 22018A gene variants.

However, in further confirmation of the fact that Africans carry considerably more gene variants than non-Africans, the variant found to provide the lactose tolerance function in Sudanese was

LCT C/G-13907, in Kenyans *LCT* T/G13915, and in Tanzanians *LCT* G/C-14010. If this population-specific character of disease-triggering variants extended beyond lactose tolerance, many if not most of the breakthroughs that were being made in genomic medicine might in fact be limited to populations of European ancestry or those with an admixture of that DNA.

Since what appears to be the same disease can be triggered by different sets of variants in different genetic groupings, breakthroughs made in identifying disease-generating variants in Americans of European ancestry were not necessarily transferable to individuals with ancestors from Africa, Asia, or even Latin America. Nevertheless, it is true that admixture ratios could make a difference. That is, if a particular African American's genomic profile is 50 percent European/50 percent African, disease triggers known to be associated with European and African ancestry can be examined for diagnostic purposes. But a bothersome problem arises in cases of admixture: as of 2021, our nation's genomic medicine industry had sequenced thirty times more disease-triggering variants in Whites than in Blacks.

The relative homogeneity of the European genome compared with the heterogeneity of the African genome may have, at least initially, led medical researchers to believe that breakthroughs being made in precision medicine and oncology had universal applicability. But such was not the case. However, the medical community, in its zeal to commercialize new genomic products, has been slow to acknowledge its initial missteps.

In short, the massive genetic databases generated as a result of the HGP did not identify genotypes that could be applied to everyone. Currently, the databases that work for Whites are unable to differentiate neutral variants from those that generate medical disorders in non-European populations. Thus, the attempt of medical specialists to identify the link between any single set of non-European gene variants and disease causation is merely a guesstimate with a thousand-to-one chance of being correct.

EXPLOITING BLACK BODIES

The precision-medicine community appeared to have been caught off guard when in November 2018 an article entitled "Assembly of a Pan-Genome from Deep Sequencing of 910 Humans of African Descent" was published.[23] A team from Johns Hopkins University, led by computational biologist Steven Salzberg, had sequenced the genomes of nearly 1,000 African Americans and found that 296,485,284 (or 10 percent) of this group's base pairs were missing from the HGP's reference genome. In a subsequent interview, Salzberg affirmed, "Eighteen years after finishing the human genome, why are we still relying on just one genome, a mosaic of a few dozen people, to guide thousands of experiments? We can do far better."[24]

In fact, even the risk loci used to match diseases to possible gene variant triggers in genomic therapies across the country have been skewed to the disadvantage of minorities. In the case of colorectal cancer, from which African Americans suffer unusually high rates, a 2017 study reported that all of the risk loci identified through the use of GWAS were found to be linked to patients of European descent.[25]

Cell-line errors were even more numerous. According to a research study by Stanley E. Hooker Jr. and colleagues, cell lines marketed as European tend to be 97 percent accurate. The study included these findings: "For instance, the 22Rv1 prostate cancer cell line was recently found to carry mixed genetic ancestry using a much smaller panel of markers. However, our more comprehensive analysis determined the 22Rv1 cell line carries 99 percent EUR ancestry. Most notably, the E006AA-hT prostate cancer cell line, classified as African American, was found to carry 92 percent EUR ancestry. We also determined the MDA-MB-468 breast cancer cell line carries 23 percent NA ancestry, suggesting possible Afro-Hispanic/Latina ancestry."[26] This degree of error will invalidate the results of any and all studies conducted using the aforementioned cell lines, and tens of millions of dollars may have been lost as the result of such flawed research.

However, when the United States of America was founded, the institution of slavery was not merely a profit-generating juggernaut built on unpaid labor; it was also a boon to the nascent field of medical science. Rather than exhume bodies from graveyards for experimentation (as physicians did in Europe during the medieval period), American medical schools purchased sick slaves for the hands-on training of students of human anatomy. Bioethicist Harriet Washington's 2007 book details the experiments of celebrated surgeon J. Marion Sims, known for generations as the "Father of Modern Gynecology." In the 1840s, Sims exclusively used Black infants for dangerous experiments exploring tetany (a children's neuromuscular disease characterized by convulsions and muscle spasms). While later medical researchers linked the disease to chronic malnutrition, Dr. Sims erroneously believed that it was caused by the displacement of skull bones during birth. According to Washington, "[Dr. Sims] took a sick black baby from its mother, made incisions in its scalp, then wielded a cobbler's tool to pry the skull bones into new positions. . . . Sims' attempt to 'open' the skull was based upon a scientific myth that the bones of black infants' skulls, unlike white infants', grew together quickly, leaving the brain no space to grow and develop." Washington then added: "This [so-called] premature closing of the black skull was held to cause low intelligence and perpetual childishness in adult blacks. When the infants died, Sims castigated the sloth and ignorance of their mothers and the black midwives who attended them."[27]

The nineteenth-century gynecologist became best known for developing surgical procedures to treat vesicovaginal fistulas, a common childbirth complication at the time in which urine from the bladder was involuntarily discharged into the vagina. He used female slaves as anaesthetized clinical trial subjects in step-by-step testing of ether use, which, consequently, became standard medical procedure for that era—but only for White women.[28]

Washington's book also discusses the more recent history of appalling medical experimentation on African Americans, including the Tuskegee Syphilis Experiment (1932–1972), a study requiring that African Americans with the disease be denied treatment in order to track syphilis's natural progression—which included blindness, insanity, transmission to partners and unborn children, and death of participants/subjects.

We must keep this narrative of exploitation in mind when contemporary medical researchers cite the paranoia of underrepresented groups and use it as an excuse for the medical establishment's failure to include such communities in clinical trials. In addition to the unhealed scars resulting from the experience of unethical experimentation, we must also note how seldom the results of medical studies that have used minorities as subjects are shared with or concretely benefit the participants. When the findings of such studies have produced genuine breakthroughs, medical researchers all too often apply them to the more profitable White healthcare markets.

FROM BIOPIRACY TO NEGLECT

In reflecting on America's sordid history of unethical use of Black bodies, it is important to note that times have changed. The twenty-first-century genetic marketplace has far less need for bio samples from marginalized Blacks, Native Americans, or Latinos. Previously, bodies of minorities were exploited because they served as inexpensive objects of experimentation for improving the health of Whites. In truth, today's precision medicine environment is far more likely to toss out the biodata of marginalized groups—as homogeneity and precision in the genomic profile of groups reduces by millions or even billions of DNA markers the number of variables that researchers must sift through to match a particular group's disease risk with the triggering gene variants. In fact, the variant matches being sought for an American of English ancestry will in all likelihood require a level of precision that even discards those belonging to a German person in favor of another member of the English ethnolinguistic group. Thus, variants of even more genetically distant populations (such as Mexicans, African Americans, or Native Americans) will be seen as little more than contaminants.

Even so, stratification efforts such as those that draw attention to sickle cell anemia in Blacks have created opportunities for discrimination and racial abuse. Physicians have not always understood the difference between the asymptomatic sickle-cell trait carried by 1 in 10 Blacks and the actual disease. In the past, employers seized upon the so-called disabling Black disease as a way to discriminate against African American employees and, in some cases, completely eliminate them from their workforces. Other demographic groups from malarial regions of Europe and India were spared such harassment because sickle cell anemia was labeled a "racial" disease. Health and life insurance companies as well as the US Air Force Academy began disqualifying Blacks.[29] Over the years, these worst-case practices have mostly been eliminated through better educational training and counseling. But as recently as 2020, cases have emerged where the deaths of young Blacks in police custody have been attributed to the sickle-cell trait.[30]

For the first time in American history, the latest tools of genetic ancestry testing can eliminate the arbitrariness and discriminatory impulses that have plagued sickle-cell testing. Ancestry testing also helps to correct some of the most pathological aspects of slavery's legacy. For instance, recent DNA testing has from time to time identified blonde, blue-eyed "Whites" who carry a percentage of West African DNA sufficient to put them at risk for sickle cell anemia. That piece of genetic knowledge would potentially have escaped diagnostic notice earlier in this country's history because few White individuals would have dared include a Black ancestor in their medical history. (In the antebellum South, the taint of having African blood, in addition to being shameful, would have relegated a previously free person to perpetual servitude.)

THE AFRICAN AMERICAN GENOMIC CHALLENGE

Even given the growing use of terms like "diversity and inclusion" in the field of genomics, there is no African American reference genome with medical benefits to Blacks comparable to that found with GWAS analyses of disease-matching gene variants in Americans of European descent. This is not to say that researchers are consciously overlooking this inequity. But the complexity of this issue is so deeply intertwined with the roots of slavery that medical studies seem almost hesitant to acknowledge the problem. As we know from current DNA evidence, the ancestors of African Americans of slave descent are West African farmers who inhabited the interior of the continent and inland areas of the Senegambia region.[31] Given the sexual nature of what textbooks euphemistically refer to as "chattel slavery," they are an admixed population. The mean DNA distribution is 75 percent West African/24 percent Northern European/1 percent Native American. However, scientists now recognize an unanticipated and counterintuitive truth: African populations, even though they share similar outward traits, carry the largest amounts of genetic diversity. Slave traders kidnapped the ancestors of African Americans from myriad ethnolinguistic groups that exhibited greater genetic diversity with one another than with a typical European (because the genetic history of Europe was generated by a subset of the African parental genome). To complicate matters further, antebellum plantation owners avoided purchasing slaves from the same ethnicities in hopes of avoiding rebellions organized by those who spoke the same languages. The long and short of it is that, regardless of the many GWAS and polygenic risk studies that have been and are today being conducted on Blacks, the breadth of the community's genetic diversity is yet unknown—though it is clearly larger than that found in the Eurasian subsets. The less homogeneous an ethnic grouping, the more challenging it is to match disease-triggering gene variants to the medical conditions they generate. And we are compelled, once again, to acknowledge that if less than 3 percent of the African genome has been studied and sequenced as of 2022, then precision and genomic medicines have a lot of catching up to do. It is far from clear whether sufficient data exist to constitute an "African American genome" within the limitations of current genomic research. There are in excess of 2,000 ethnolinguistic groups on the African continent compared with 87 in Europe.[32] Because inhabitants of Africa retained the full range of gene variants that sculpted us into modern humans, each ethnolinguistic group will carry at least as many gene variants as the European groups in aggregate. Even though the Danish genome differs from the Irish genome, the disease-triggering gene variants causing many illnesses in both groups will overlap. But that will often not be the case with African Americans because the history of the antebellum South has artificially lumped into one ethnolinguistic group hundreds of enslaved West African ethnicities. Given that genetic diversity is still being interpreted in many academic circles as the aggregating into a single pool as many minority groups as can be found, it is unclear how such a mish-mash will accomplish the gene-variant matching that is now successfully taking place in therapies involving Americans of European descent. The United States has recently developed massive big-data computer systems. And yet, however impressive their capacity to analyze large datasets, these super computers cannot generate valid hypotheses in a knowledge vacuum. For purposes of medical accuracy, the United States may need to rely on diplomacy to secure reference genomes from nations around the world (particularly those whose populations are more genetically homogeneous and thus more easily matched to their own unique disease-triggering variants).

Despite these challenges, the two most commonly identified gene variants in American Blacks—the African variant of the TRPV6 calcium ion channel and the Apolipoprotein1 G1 and G2 variants—will undoubtedly play a role in closing the racial health disparities gap. But currently, the medical-research community continues to overlook disease triggers that are not linked to illnesses in populations of European ancestry. This issue is also exacerbated by the difficulties associated with developing taxonomic systems that effectively differentiate African ethnolinguistic groups and communities in the diaspora such as African Americans of slave descent and Afro-Caribbeans. For instance, a recent classification model being used in research studies references "Atlantic Africans"—a term used to designate victims of the trans-Atlantic slave trade. But for research related specifically to high rates of hypertension, kidney failure, and cardiovascular disease in African Americans, this taxonomy does not differentiate Africans who, as coastal dwellers, were genetically adapted to twenty-five times more sodium intake than those from the interior of the continent.

POLYGENIC RISK SCORES

When aggregating genomic data from hundreds of thousands of individuals, scientists over time noticed that complex diseases often involved large numbers of variants, each of which minimally affect symptomatic outcomes. In 2018, a team led by Amit V. Khera of the Broad Institute in Cambridge, Massachusetts, coined the term Polygenic Risk Scores (PRSs).[33] The methodology, initially used for selective breeding in plants and animals, provided a means of summarizing the collective effect of different gene variants on a single organism. It was first applied to humans in the late 2000s as a means of identifying individuals at high risk for disease. Using GWAS techniques to assess genetic variations associated with disorders—including heart disease, atrial fibrillation, type 2 diabetes, inflammatory bowel disease, and breast cancer—the team developed a computational algorithm that combined the variants into a number (or PRS). The scores offered more cost-effective ways of determining the risk of a particular disease at the individual level and could be used, according to the algorithm's creators, to determine a person's inherited susceptibilities to the disease.

Even though PRSs were being hailed as one of the most important breakthroughs in genomic medicine, critics noted that the use of PRS-generated data was creating a vast chasm of inequality for people with ancestors outside of Europe. Massive amounts of data for Whites were already available in the database and even more were added daily, so the result was more effective matching of gene variants to specific diseases within that particular defined population. But equivalent amounts of data were not being aggregated for many populations defined as non-European. (To be fair, the challenges and, therefore, costs of doing so may have seemed prohibitive compared with the significant amounts of data that had already accumulated on a single defined population.)

Not surprisingly, European-derived PRSs were poor predictors of disease susceptibility in American minorities of non-European genetic ancestry. For this reason, Alicia Martin, a genome scientist at Massachusetts General Hospital, once quipped, "the Polygenic Risk

Scores for people tracing their ancestry to Africa were only marginally better, if at all, than flipping a coin."[34]

In response to the absence of underlying data on minority groups, research institutions have been recruiting more subjects from these populations. But to be effective, the design of such studies must identify populations in genomically relevant ways (that is, they must be stratified). Populations that share similar sets of disease-triggering gene variants because they come from similar ancestral population niches need to be grouped together.

Given the history of exploitation, some minorities may be afraid of exposing themselves to genetic experimentation. But for most who yearn for any legitimate medical breakthroughs that might ease their pain or disease symptoms, it so happens that non-European gene variants continue to be eschewed by researchers as being of little scientific value. But given the public emphasis on issues of diversity, how could this be? The challenge for scientists in this age of GWAS is not trying to identify or name previously unnoticed gene variants but rather trying to figure out what these genetic alleles do (that is, what role a variant plays in the body of a human). Is it disease-triggering, neutral, or functional in some evolutionarily positive way? Without closely observing the habits, habitat, disease risks, historical migratory patterns, and cultural patterns of the human population in question, scientists can merely make guesstimates. In most cases, the less known about the group the less accurate will be the guesstimate, regardless of the sophistication of the scientific methodology being applied.

It was initially believed that GWAS would help scientists figure out what gene variants do. But they are able to do so only when the genetic population under study is sufficiently homogenized and their cultural habits as well as ecological landscape well known. Such cases must involve large numbers of individuals who happen to share the same sets of gene variants and whose cultural behavior patterns are familiar to the researchers. Because of the population-specific (but not racial) nature of disease-triggering gene variants, those found outside of a narrowly defined but large sample population will not be matchable because the genetic possibilities can be exponential.

A more sophisticated use of arithmetic manipulation reminiscent of eugenics can be found in the book *The Genetic Lottery: Why DNA Matters for Social Equality* by behavioral geneticist Kathryn Paige Harden. In the book, Harden coined the term "anti-eugenics" (perhaps in an attempt to differentiate her views from ongoing controversies regarding cognitive hierarchies based on race). But perhaps, in more subtle ways, Harden's work reinforced the long-held notion among race scientists that genes play a pivotal role in educational achievement, income levels, and social inequality in the United States. In the Preface, she notes that her research data are restricted to Whites. But nowhere else is this distinction made, leading readers to surmise that the author's references to race and genetic factors in educational achievement and socioeconomic levels are backed by data.[35]

Once the methodological gates had been pried open by Harden's book, new polygenic tools were soon adopted in fields as disparate as genetics and education—where PRSs were being used as definitive proof of a human cognitive hierarchy. Not surprisingly, controversy ensued. A 2021 article asserted, "Recently, the author claimed polygenic scores provide evidence that a significant portion of differences in cognitive performance between Black and White populations are caused by genetic differences due to natural selection, the 'hereditarian hypothesis.'" The authors concluded, "Cognitive performance does not appear to have been under diversifying selection in Europeans and Africans. In the absence of diversifying selection, the best-case estimate for genetic contributions to group differences in cognitive performance is substantially smaller than hereditarians claim and is consistent with genetic differences contributing little to the Black-White gap."[36]

In fact, as recently as 2019, scientists at Stanford and Harvard began questioning the efficacy of polygenic risk scoring when used beyond the parameters of European-defined populations. Their findings warned that the predictive value of polygenic scores derived from European ancestry was only one-third as informative for African ancestry. That level of difference may be even more disappointing than the researchers recognized. When we account for admixture ratios, it turns out that, on average, one-quarter of African American DNA is Northern European. Might that mean that such analyses are capturing merely one-twelfth the predictive value of Blacks' African DNA?[37]

The concept of using polygenic risk scoring for "educational attainment" polemicized a previously neutral descriptor that had been used in the US census to define "the highest level of education that an individual has completed."[38] But educational specialists were now asserting that PRSs could be used to predict which kids would succeed in school. Such scores had come to fill the gaping hole that had been left by the growing stigmatization of Intelligence Quotient (IQ) tests as racially biased. A 2022 study on the use of polygenic prediction of educational attainment involving three million individuals offered the following clarification: "PGIs like ours that are constructed from GWAS in samples of European genetic ancestries are generally found to have much lower predictive power in samples with other genetic ancestries; for example, on average across phenotypes, estimates of relative accuracy (ratio of R^2) in African-genetic-ancestry to European-genetic-ancestry samples have been 22 percent and 36 percent."[39]

Here we need to pause and question the results as well. Any genetic testing on African Americans that identifies an efficacy rate in the 25 percent range can be dangerously misleading. This population is primarily an admixture of West Africans and Northern Europeans, with the latter's ratio being 25 percent. One cannot know without testing of ancestral ratios whether the 25 percent predictive power might be solely a function of that cohort's European DNA, which superficially appears to validate the whole of the test as regards Blacks.

RACE AND EUGENICS

By the twenty-first century, the scientific community should have demolished the pseudo-science of eugenics and applied the latest discoveries and intriguing new insights that have emerged from the HGP. After all, a definitive understanding has been reached among both scientists and academics in the social sciences that race is a socially constructed, unnatural classification scheme used to differentiate and usually exploit other humans. As for why so little has changed, we need only look at the proportion of the US electorate that endorses pro-eugenics politicians to undermine foundational principles of democracy. We might even look at their wealthy financiers. In short, there is far too energetic and well-financed a market for racist pseudo-science to allow it to disappear merely on account of its own contradictions. Traditionalists who have clung to an antiquated notion of race continue to see nothing wrong with grouping together humans with similar physical traits because the assumption is that these differentiations would be reflected in their genes. But the scientific conception of human genetic diversity contradicts this "essentialist" concept of race. And yet "essentialism" remains as omnipresent in twenty-first-century debates as it was in our nation's turbulent racial past.

Before the emergence of fields such as genetics, molecular biology, and genomics, all we humans had to go on in distinguishing one another were physical differences. This essentialist notion was accompanied by a great deal of baggage—insecurities and projections—and assumptions, chief among them being the belief that because of Africa's underdevelopment, its inhabitants were isolated tribes living on the margins of human evolution.

Why, indeed, should we expect science to efface eugenics overnight when science gave birth to it? As early as 1839, Professor Samuel George Morton, one of the founders of physical anthropology, asserted that he could discern human intelligence through skull measurement, thus pioneering the pseudo-scientific field of craniometry. Theories relating darker skin color to a plethora of dehumanizing traits are as old as American slavery itself. In fact, the legal institutionalization of slavery beginning in the mid-seventeenth century carried along with it a necessary rationalization. European monarchs found Americans' self-righteous proclamation of all men being created equal in some ways ludicrous. No slave plantations existed in the British Isles or elsewhere on the European continent. But what if, as Americans began to assert, the enslaved West African farmers were inferior beings who would be uplifted by the traditions and structures of slavery? In fact, one of America's founding fathers, Benjamen Rush (1745–1813) insisted that black skin was a hereditary disease. US founding father and slave owner Thomas Jefferson (1743–1826), who fathered six children with his Black mistress Sally Hemings, was even seen at the time, according to microbiologist Joseph L. Graves, as "one of the most influential pre-Darwinian racial theorists."[40] *The Types of Mankind*, written in 1854, contributed to this burgeoning discussion by suggesting that "Negroes" held a rank in the human hierarchy somewhere between "Greeks" and chimpanzees.[41] British natural scientist Francis Galton was the first to coin the term "eugenics," which further legitimized racial hierarchies. Galton asserted that "the more suitable races or strains of blood (were given) a better chance of prevailing speedily over the less suitable than they otherwise would have had."[42] Even though Charles Darwin proposed the hypothesis of a single origin of humans, he nonetheless referred to the "savage races" such as "the negro or the Australian" as being closer to gorillas than were White Caucasians.[43] But it was Galton's concept of using precise measurements as a means of proving the scientific nature of African inferiority that gained even greater momentum in American universities at the turn of the twentieth century. Controversy continued to swirl around Darwin's theory that all humans had originated from the African continent. In 1912, amateur archaeologist Charles Dawson injected into this discussion a shocking new discovery.[44] He presented the so-called discovery of the Piltdown Man as evidence that an older European lineage of homo sapiens had inhabited the British Isles. Over the years, several skeptics had questioned this theory. But it was not until 1954 that William L. Strauss published the definitive proof that the Piltdown Man was a hoax, and that the mandible of a primate had been glued to a human forehead.[45] By then, the notion of linking Blacks to subhuman standards of intelligence merely morphed into a different set of topologies. No longer were skull measurements used for this purpose when IQ tests were so much easier to grade and use as measuring instruments.

The eugenics movement reached its height in Nazi Germany during World War II and encouraged the use of "planned breeding" for purposes of racial improvement. A coterie of American "race" scientists helped to shape this ideology into a presumed science. The concept's fallacy was monumental. But it could only be perceived in light of advancements in genetics that would not mature until the beginning of the twenty-first century. The breeding of so-called superior human stock, rather than leading to the desired improvements, led to genetic autosomal diseases caused by inbreeding. Commercially savvy dog breeders may have convinced some members of the general public to pay exorbitant sums for dogs marketed as "elite" breeds. However, the selective breeding of canines, like that of any other living organism, is a highly subjective function of whimsy or taste. The farmer's or breeder's sense of what will bring market value differs from the calculation applied to living organisms undergoing involuntary genetic manipulation. This weeding out of the so-called inferior stock has even been given a name in history textbooks—the Holocaust—and wars of ethnic cleansing have occurred across the globe.

The eugenics movement thrived in early twentieth-century America, where the legacy of slavery created an imperative to position Blacks and Whites in a biological hierarchy.[46] As early as 1902, Stanford President David Starr Jordan was popularizing the concept of "race and blood" according to which the supremacy of Whites was transmitted through their blood.[47] Some scholars assert that the vitality of the US eugenics movement spread to Germany in the 1930s.[48] However, the ideology lost its fervor after World War II as few Americans were willing to associate themselves with the now-discredited theories of Aryan superiority advocated by Nazi Germany. As a result, the movement went underground, largely lying dormant to later resurface with a vengeance in American academia. In 2001, the *Albany Law Review* published an article by Paul A. Lombardo detailing the racist roots of the Pioneer Fund.[49] Not long after, a book by William H. Tucker appeared on the use and abuse of science in support of bigotry.[50]

Contrary to Adolf Hitler's notions of racial superiority through Aryan inbreeding, modern science has shown us that the more genetic distance that exists between two reproducing members of the same species, the more biologically vigorous their offspring will be. This concept—referred to as heterosis or hybrid vigor—is a reminder that the survivability of any biological species will depend on having the broadest range of genetic variation. But reality does not negate the dark and fierce impulse embedded within the competitive nature of humans to perfect their progeny in whatever way deemed most aesthetically desirable at a specific moment in history.

The 1994 publication of *The Bell Curve* by Harvard psychologist Richard Herrnstein and political scientist Charles Murray reinvigorated interest in presumed cognitive differences among the races.[51] J. Philippe Rushton, a Canadian psychologist, insisted that Whites had higher IQs, reasoning, "It's a trade-off, more brains or more penis. You can't have everything."[52] As nonsensical as this assertion may sound, it at the very least exposes the underlying preoccupation of *The Bell Curve*'s authors with the National Basketball Association (NBA). The book's arguments regarding the lower cognitive range of Blacks references the NBA numerous times, concluding that just as muscularity is biological, the fact that Blacks score lower on IQ tests is merely another unbiased genetic measurement.

The 1992 sports comedy *White Men Can't Jump* about two street hustlers added yet another measurement to a masculinity competition in which Blacks were not even informed that they were central players. Nevertheless, how high a male can jump and his prowess in basketball became more cultural grist for the eugenics mill. Several years later, Vincent Sarich, a professor at the University of California–Berkeley and supporter of psychologist Arthur Jensen's racial theories, gave a noted lecture, later cited in *Science,* in which he asserted, "If you can believe that individuals of recent African ancestry are not genetically advantaged over those of European and Asian ancestry in certain athletic endeavors, then you probably could be led to believe just about anything. But such dominance will never convince those whose minds are made up that genetics plays no role in shaping the racial patterns we see in sports."[53]

The race-oriented literature of that era defended itself against being tagged racist by placing Asians above Whites in the cognitive hierarchy. However, upon closer inspection, the fact that the new eugenics literature presented Asian males as having high IQs but being undersexed and thus weak rivals revealed the consistency of its masculine preoccupations.[54] Americans of slave descent are an admixed population sharing both West African and Northern European DNA. The term "Black" is what academics refer to as a construction. Such terminology, although it appears descriptive of nature's biological forces at work, is a human contrivance. Why was it necessary for slave owners to define "Black" in terms of the mother's status or lineage in rigidly patriarchal colonial America? Slave owners who sexually assaulted female slaves on their plantations designated the resultant offspring as "Black" regardless of the ratio of Northern European to African traits the child had inherited. The number of slaves on a plantation was dramatically increased in this way. While every other natural phenomenon on these massive farms, from crops to cattle husbandry, was understood to become more vigorous from cross-breeding, the derogatory term "mongrel" was associated with the mixing of human DNA from the two genetically distant populations in the South.

David Epstein, author of *The Sports Gene,* saw through the confusion caused by usage of "Black" in human classification and noted that "with only ninety thousand years for unique changes to occur outside of Africa, there simply hasn't been as much action in many stretches of the genome. People outside of Africa are descendants of genetic subsets of a group that was itself just a subset in Africa in the recent past."[55]

In the antebellum South, any person suspected to be of West African ancestry would be condemned to perpetual servitude—whether they had a pale complexion, blonde hair, or blue eyes. What no one could see at the time was that the most enduring legacy of this classification scheme would be the diminished self-regard that a segment of the White male population would develop during slavery and carry around to this day. It is for this reason that so much eugenics-oriented material presents images and metaphors related to Black muscularity, penis size, and athleticism.

Because sexual exploitation was integral to slavery, we would today be hard pressed to find in the antebellum South African Americans descended from slaves who were not genetically admixed with

Northern Europeans. According to K. D. Zimmerman, even the relatively isolated Gullah Geechee communities of South Carolina possess some English and Irish DNA, albeit less than the US Black community as a whole.[56] Such would not necessarily be the case with post-slavery immigrants from the Caribbean or Africa.[57]

Admixture of distant defined populations within the same species carries biological advantages. In fact, the term "hybrid vigor" (also known as heterosis) describes the enhancement of specific biological traits in a hybrid offspring over its parents. In humans, this effect can be seen from the mating of two individuals from genetic populations that are not adjacent to one another whose children will be taller than their parents.[58] The simple fact is that Blacks receive all the credit for the hybrid vigor of an athleticism that emanates from a two-stage diversifying process. The European equivalents of the American racial term "White" are the 87 distinct ethnolinguistic groups who generally marry within their distinctive groupings (such as Germans marrying German speakers, English marrying English speakers, Italians marrying Italian speakers, and so on). However, once they arrive in the United States, the very fact that Whites tend to marry other Whites creates a veritable explosion of genetic diversification. Because African Americans were brought to the United States as enslaved persons/farmers, they aggregated the gene-variant pools of an even larger number of ethnolinguistic groups (such as Yoruba, Fon, and Malinke). Adding to this exceptional pool of genetic variation is the fact that African Americans are the product of the genetic diversification of both European and West African ethnolinguistic populations. However, to the degree that human exceptionalism has any meaning at all, it has nothing to do with genes. Once we unburden ourselves of the notion that whatever community "I" belong to is superior to everyone else's, then we see the light. The collective genius of human creativity is the gift that will be bestowed on whatever nation figures out how to detribalize its citizens and check their impulse to perceive hierarchies where they do not exist.

HUMAN HISTORY DEFLATES HIERARCHICAL THINKING

Eugenicists once used differences in skull shapes and sizes to measure human intelligence, which relegated Africans to the bottom and Europeans to the top. Research has proven that the mid-nineteenth-century theory of skull size, weight, and shape is false. Humans with the largest as well as the smallest craniums, and all skull sizes and shapes in between, are found on the African continent.[59] In fact, whatever brilliant characteristics one race claims to have (such as skull size and superior intellect), the reality is that these cognitive traits were inherited from African ancestors and are to this day carried by some individuals in Africa. Since twice as many gene variants will be found in Africa as anywhere else on the globe, who knows what combinations of Einstein, Shostakovich, and Madame Curie might be lying dormant in the African parental genome—or more likely carried by an impoverished subsistence farmer in Zimbabwe.

The only challenge with finding Africa's Einstein is that we simply may not recognize a genius the instant we come upon one. Not all cognitively gifted people are enrolled in mathematics, physics, or computer science programs or hold professorships at prestigious universities. A Galileo or a Christopher Columbus born into a rural, subsistence society might live in a mud hut, have a phenomenal memory for the particulars of the family's goat herds and a compass-like sense of direction, but no time between chores to contemplate the true nature of the universe. But we as Westerners have woven into our collective ego so many mythologies about our own exceptionalism that some of the most fundamental lessons of human history have yet to penetrate that barrier. The fallacy that has tripped up eugenicists and non-eugenicists alike throughout history is the mistaken belief that intelligence determines a society's level of material culture. In fact, a popular twenty-first-century subfield of eugenics is represented by academic articles linking high IQs to a nation's level of economic development. In 2001, Richard Lynn and Tatu Vanhanen published an article in which they asserted, "Because national IQs are substantially correlated with per capita income, it can be assumed that national IQs must have been associated with economic growth at some time in the past. We showed that this was the case over the 500 years from 1500 to 2000 for which IQs for 109 nations were correlated with rates of economic growth at .71."[60]

But what accounts for the fact that some societies build pyramids while others make do with mud huts? We as Westerners at times wear blinders and simply refuse to acknowledge that the progeny of the pyramid builders today live in mud huts. The notion that humans inhabit a progressive universe is the essence of naivete. Some of the most haunting passages in human history are the laments of privileged Egyptian scribes, Greek poets, jurisconsults of Timbuktu, and chroniclers of other great civilizations who had no way to see the end coming until it collapsed on top of them. And even though military history books entertain us with the details of great battles, ecological disaster is what often undermines ancient empires from within long before their predatory neighbors show up. Ancient Egypt was worn down by the irregular flooding patterns of the Nile long before the Greeks and Romans appeared at the harbor of Râ-Kedet. Modern agricultural technology is no more able to increase surpluses in the Sahel now being devoured by the climate-energized Sahara Desert than were my West African ancestors, who were rounded up by slave-catchers.

As science writer Angela Saini noted in her book, racism still lingers in modern medical science.[61] It can from time to time be found in peer-reviewed journals, particularly when the use of technical jargon obscures its underlying theme and flawed methodology. But this is not to say that the pretense of colorblindness is the solution. Not mentioning race all too often means universalizing European genotypes rather than segmenting populations according to the specific gene variants under examination.

A December 2017 article by Alexander P. Burgoyne and David Z. Hambrick analyzed the results of a study released earlier that year conducted by Dutch researchers involving DNA sequences from 78,308 people. Their aim was to determine whether correlations existed between specific gene variants and intelligence. Burgoyne and Hambrick concluded, "As a check on the replicability of their results, the scientists then tested for correlations between the 336 gene variants and level of education—a variable known to be strongly correlated with intelligence—in an independent sample of nearly 200,000 people who had previously undergone DNA testing. Ninety-nine percent of the time, the SNPs correlated in the same direction with education as they did with intelligence. This finding helps to allay concerns that the SNP's associated with intelligence were false positives—in other words, caused by chance."[62] The authors failed, however, to include an important caveat mentioned in the original study—that the genome-wide association meta-analysis had been conducted solely on individuals of European descent.[63] Rather, the study language assumed universal applicability of the research findings. Because the target population is not mentioned, members of the American scientific community—likely the principal audience for such an article—may subconsciously apply these findings to the educational disparities constantly being referred to in the media. After all, the entirety of both public and scientific attention paid to supposed hierarchies in cognitive ability pits Whites against Blacks. The issue of whether the low levels of intelligence of some Whites as measured on IQ tests correlate to certain gene variants found in the European genome is not, and has never been, posed. Rather, in the United States, any findings of differences will be instinctively applied to the question of whether Blacks score lower on IQ tests than Whites because of genetic rather than socioeconomically determined differences.

A 2020 article authored by Burgoyne and colleagues purported to show a link between human intelligence and baseline pupil size. While the supporting data were not included in the article, they were made available in the *Mankind Quarterly*, a journal that has often been referred to as the "cornerstone of the scientific racism establishment." Data omitted from the *Scientific American* article that appeared in the *Mankind Quarterly* reported that mean pupil size was 3.56, 3.35, and 3.23 for Whites, Hispanics, and Blacks, respectively. The authors explained that these findings placed Whites at the top of the cognitive hierarchy as a function of "evolution and 'geo-bio-climatic' selection" (whatever that means).[64]

Recently, we've seen the real-world implications—outside of print media and in headlines—of modern-day eugenicist beliefs. In May 2022, an eighteen-year-old shooter motivated by the racist writings of a British eugenicist named Michael Woodley killed ten Black people in Buffalo, New York.[65]

EXPANDING THE GENETIC LANDSCAPE

When we take a larger view of the positioning of African Americans in genomics research and include the subfields of genetic architecture, epigenetics, and meta-genomics, the picture remains formulaic. A 2011 article entitled "Genetic Architecture of Cancer and Other Complex Diseases: Lessons Learned and Future Directions" announced, "In Europeans, the increase in genomic coverage and accumulation of large numbers of densely genotyped samples for many common diseases will reach a point where only variants with very small effects or very low allele frequencies remain to be discovered. In addition to pursuing dense genotyping and sequencing in these populations, an important opportunity to pursue now is to extend discovery efforts to underrepresented populations. European ancestry populations have less genetic diversity than African ancestry populations and patterns of linkage disequilibrium are known to vary by population."[66]

Ten years later, the only evidence of progress found in studies on the genetic architecture of Blacks appears to be merely an expanded profile of diseases in which the triggering gene variants differ from those of Whites. A cursory review of the medical literature will show that this list of diseases now includes inflammatory bowel syndrome, asthma, type 2 diabetes, prostate cancer, hypertension, kidney disease, and cardiometabolic disease. The only course of action consistently articulated in journal articles is the need for further research on the genetic architecture of African American disease-triggering variants.

It has been gratifying in recent years to see the ways in which the medical world has owned up to the wrecking ball effect that our nation's history of racism has had on Black health and well-being. But what has also crept into these kinds of discussions and research initiatives is a kind of scientific sloth. Hypertension in Black Americans of slave descent is not caused solely by stress related to racism. Rather, the cause is the intake of a toxic level of sodium found in the American diet that, while high, is not fatal to mainstream America. Let's not turn those few racial disparities that have practical solutions into cosmological issues of race and equity that have still not been resolved two-and-a-half centuries after our nation's founding.

FINANCIAL DISINCENTIVES

By 2025, the global precision medicine market—a largely for-profit venture based on the original data in GWAS—is projected to reach $3.18 trillion with a compound annual growth rate of 10.6 percent.

Medical capitalism is thriving.[67] American clinics are charging upwards of $400,000 per person for cancer genome sequencing—a procedure that could be health-saving if one has primarily European DNA. After all, the more uniform the variants found in the patient's gene, the higher the chances of successfully identifying disease-triggering variants by comparing them with others of the same ancestry. Eighty-eight percent of the databases used in genomic analyses of disease causation are populated with data on Europeans. The DNA of African Americans, by contrast, is on average 25 percent Northern European. While GWAS could analyze with accuracy the European portion of the admixture ratio, it has not been able to draw accurate medical conclusions because the African portion—75 percent—is unknown.

The fields of genomics and precision medicine monetized their science before assessing the medical parameters and limitations. Widespread media coverage added to the missteps. If we go back and look at nearly every headline announcing a medical breakthrough after 2010, we can clearly see that the genetic research and clinical trials were performed primarily on Americans of European ancestry, who reaped the benefits—the cures and therapies that worked solely on their specific, defined population.

Science reporters failed to recognize or communicate to the American public how rigidly and narrowly focused the market was for each new discovery in medical genomics. They also failed to point out that the individualized nature of the new precision medicine only worked for individuals within a well-defined population; that is, the data were not interchangeable between ethnicities or demographic groups. Personalized medicine was and is not for every person. While correcting the problem politically by "diversifying" our reference genome may sound like progress, as long as it is not matching similar disease-triggering gene variants to specific populations, its benefit to minorities will remain inconsequential.

3 OUR HEALTH

TRACKING DOWN ANCESTRAL CLUES

EVERY SUMMER, Mama would pack my brother, sister, and me into our aging Ford station wagon for the trip from the tiny Black enclave of New Cassell on Long Island, New York, to Louisiana. Once we crossed the Mason-Dixon Line into the world of segregated bathrooms and water fountains, the pace of the 1,500-mile journey slowed and multilane highways became back roads. My mother, a feminine, graceful woman, was also a World War II veteran. With a loaded rifle under the car's back seat, she drove for three days and three nights straight, making stops only for the bathroom or to buy gas until we reached the relative safety of Mermentau, Louisiana. The marshes surrounding the tiny rice mill town were once rumored to shroud the pirate's cove of Jean Lafitte. Mother's homestead might have been slightly less segregated than other places in the South if only because the cost of maintaining residential boundaries was more than the local inhabitants could afford.

Mama's kin were mostly a jumble of Baptists, lighter-complexioned Roman Catholics, and practitioners of a religion called "Vodun" (which, when translated from the Fon language of West Africa, vaguely meant "introspection into the unknown"). The laughing, loving women would at times drop their voices to a whisper when I entered the room. What secrets were they unwilling to share with Ida Mae's nervous, delicate-natured daughter? No matter. I too had secrets. From the age of six,

I suspected that no bearded white man floating in the sky, let alone his dead son, was coming to save us. (In later years, I saw the human dilemma more clearly.) However cacophonic my relations' profusion of religious beliefs appeared on the outside, on weekends they released themselves to Afro-Caribbean cross-rhythms, Louisiana swamp blues, and zydeco. In processing raw emotion through dance, believers and non-believers alike in my family circle found themselves lifted into fleeting moments of clarity and, sometimes, joy.

Breast cancer snuffed the life out of my vivacious mother in 1977 at the age of fifty-one. Physicians had not yet given a name to the aggressive subset of the disease that disproportionately strikes down Black women in middle age. But given the cancer's aggressiveness and Mother's ethnic background, current medical lore would probably have pinpointed the disease—triple-negative breast cancer (TNBC), identified by the estrogen receptor (ER) negative, progesterone receptor (PR) negative, and human epidermal growth factor receptor 2 (HER2) negative. Mama died within months of her diagnosis.

THE TSETSE BELT AND BONE HEALTH

In the years after Mother's death, I traveled to West Africa and buried my grief in fieldwork, exploring research questions regarding the link between the chronology of historical events and epidemiology. The zone extending from the southern borders of the Sahara to the humid Atlantic coast is known as the Tsetse Belt. An infestation of the tsetse fly (*glossina*) covers this region of the continent and has done so for millennia. These vector-borne parasites that scar the landscape (trypanosomes/single-celled parasites) feed on the blood of their victims. While some species pass on sleeping sickness to humans, the *Trypanosoma brucei* strain that is most common in West Africa primarily attacks cattle with a wasting disease referred to locally as *nagana*.[1] This harsh ecological environment accounts for the fact that West African food culture is devoid of dairy products. Inhabitants of the Tsetse Belt have also been found to be 99 percent lactase nonpersistent (or lactose intolerant). Paradoxically, the available medical data from West Africa seem to suggest that osteoporosis in this region of the continent is nearly nonexistent.[2] Because the data infer strong bone health among calcium-deficient (that is, lactose intolerant) West Africans, it has been ignored in much of the medical literature and deemed illogical.

One morning in the fall of 2015, I was awakened by an idea buzzing around in my head. Although I lacked hard evidence regarding the low risk of bone diseases in West Africa, what I possessed was even more compelling—knowledge of databases regarding African American health. I reached for my computer and began searching. African Americans carry on average 75 percent West African DNA and also happen to be 75 percent lactose intolerant, which defines this community as calcium deficient by federal nutritional standards. To determine whether West Africans were susceptible to osteoporosis, I simply needed to search for rates of susceptibility to the disease in their more copiously studied African American descendants. The medical data I gathered over the course of the next several months was revealing.

An early study showed that African American adolescent girls had greater calcium retention (185+ mg. a day) and lower urinary calcium excretion than their White counterparts. The author concluded, "Adult African Americans require 300 mg. less calcium per day than do whites to replace their calcium losses from the body.[3] Another study confirmed what dozens of research reports had merely hinted at: bone mineral density (BMD) varies across race and ethnic groups, and African Americans (also known as non-Hispanic Blacks) exhibit the highest BMD in comparison to other US demographic groups. The study stated, "In conclusion, differences in bone microarchitecture and density contribute to greater estimated bone strength in

African-Americans and probably explain, at least in part, the lower fracture risk of African-American women."[4]

Given the fact that African Americans carried on average 75 percent West African DNA, their strong bones corroborated the scant data on West Africans having minimal rates of fragile-bone diseases. What fueled the ever-spiraling maelstrom of my obsession with the matter of calcium and bone health? Cancer.

TRPV6-EXPRESSING CANCERS

Calcium—99 percent of which is found in the bones and teeth—is the most abundant mineral in the human body. The remaining serum calcium is tightly regulated because of its biochemical power and capacity to wreak havoc in the body. Free calcium ions (Ca^{2+}) are involved in the contraction of muscles, absorption, and transcellular transport and have been implicated in cell proliferation and metastasis (which enables normal cells to multiply and evade cell death). Medical researchers have known for years that up to 30 percent of cancer patients will exhibit high levels of calcium (known as hypercalcemia). But the medical literature confirmed that in all cases, rising calcium levels was a side effect of cancer and not its cause.[5]

In 2016, I came across three pieces of research whose trisecting observations sent a shock down my spine. As a historian, I had neither the standing nor the qualifications to question established medical lore. But what I possessed was the capacity to identify a Eurocentric paradigm posing as a universal principle of human biology. The first two studies that captured my attention had been published in 2008. Both suggested that ancestral (African) *TRPV6*a absorbed more dietary calcium than the non-European polymorphism (variant) of that gene.[6] Those studies also showed that this variant's function appeared to be linked to the African *A563T* variant of the *TRPV5* renal epithelial calcium channel, which allowed the kidneys to retain more calcium than non-African gene variants.[7] Further research clarified that the *TRPV6*b (or "derived") haplotype carried by non-African populations was quicker to "close" in response to Ca^{2+} feedback, signifying that it would reabsorb less Ca^{2+} than the ancestral form.[8]

Earlier that year, *BoneKEy Reports* published research of mine showing that low-calcium-consuming West Africans appeared to have strong bone mineral density, which put them at low risk of osteoporosis, but I wanted to know why.[9] Might this African variant of the *TRPV6* gene account for strong bone health in a low-calcium-consuming genetic population? It was while contemplating that issue that I came across the third intersecting piece of research—an article that had been published in 2007 by French researchers that correlated the *TRPV6* calcium ion channel gene with certain aggressive cancers. N.V. Lehen'kyi and colleagues published a study suggesting that the protein produced by this calcium ion channel gene might even be a useful biomarker for determining the prognosis of a patient suffering from a "*TRPV6*-expressing cancer."[10] (That term for a specific subtype of cancer—*TRPV6*-expressing cancer—was new to me.)

But it was only after seeing the list of cancers defined as *TRPV6*-expressing that I began to suspect a biological consanguinity. Is it mere coincidence that African Americans suffer unusually high susceptibility and mortality rates from precisely those cancers defined as "*TRPV6*-expressing" (namely, TNBC and metastatic prostate cancer)? Since Blacks are found to carry the more calcium-sensitive *TRPV6*a calcium ion channel variant, might their consumption of calcium above their biological needs trigger these malignancies? Because of the critical nature of calcium in the human body and the lack of taxonomies that transcend the unscientific concept of race, medical studies seem to have steered away from exploring calcium intake in different populations. The one-size-fits-all paradigm used by the medical community

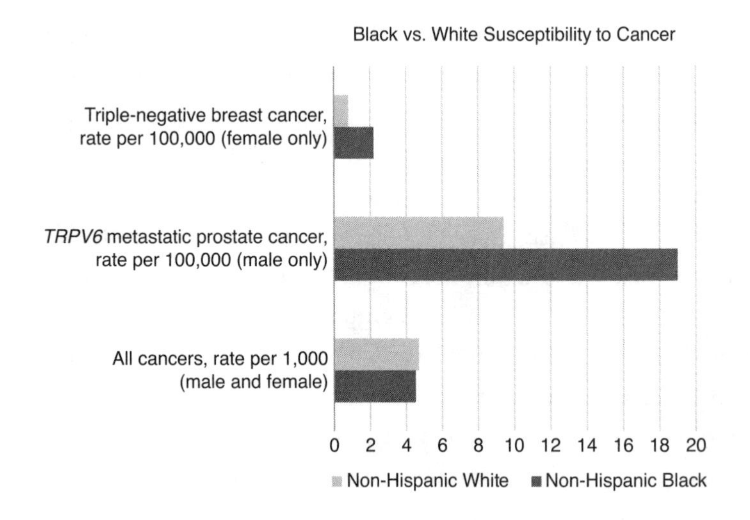

Figure 3.1 Even though Blacks generally exhibit a lower cancer risk than Whites, their susceptibility to *TRPV6*-expressing cancers is dramatically higher. *Data source:* SEER*Explorer, National Cancer Institute, Surveillance, Epidemiology, and End Results Program.

discourages researchers from questioning the possibility that a calcium-intake level in excess of biological need might trigger those cancers defined by the uncontrolled proliferation of *TRPV6*a protein. Thus far, laboratories have not noted as significant the obvious link between the uncontrolled proliferation of the African *TRPV6*a variant in such *TRPV6*-expressing cancers as TNBC and metastatic prostate cancer. (See Figure 3.1.)

The more I researched TNBC, the more questions arose regarding the role the calcium-absorbent African *TRPV6*a gene variant might play in Blacks' higher risk of *TRPV6*-expressing cancers. But the necessary data were just not available. However impressive Big Data had become over the course of the past two decades, it remained the case that gene variants not found in Americans of European ancestry were often shunted to the side and considered "rare."

The enticing fragments of knowledge that I'd accumulated left me heartsick and lost. I was not a medical researcher but an Africana historian living in a nation where certain politicians were hell-bent on erasing what few historical details of slavery were presented in American textbooks. Cancer research has understandably put in place barriers to keep out grifters and charlatans eager to make a buck on others' pain and suffering. Because the value of medical research is measured by the tens of millions of dollars allocated to laboratories and institutes by funding agencies, my research essentially had no worth because it had no price tag. So, in the silent spaces created by medical indifference, I devoted myself to piecing together the clues stacking up on my computer—and my efforts began to pay off, at least in little ways.

Several contacts in the scientific community helped me arrange to exhibit posters at the American Association for Cancer Research (AACR) conference on the correlation between calcium intake and TNBC and metastatic prostate cancer. I also published an article that called for cancer researchers to take a more serious look at the *TRPV6*a gene variant in Black Americans from an ecological standpoint (that is, Blacks inherited strong bones from their low-calcium-consuming West African ancestors and might, therefore, be maladapted to the high-calcium food environment of the United States, and the ancestral *TRPV6*a gene variants could possibly become invasive and carcinogenic when overexposed to excess free-calcium ions).[11]

My work attracted the attention of computational biologist Patricia Francis-Lyon, whose team at the University of San Francisco Health Informatics Program published an article that noted:

> We have examined Hilliard's recently proposed hypothesis of ancestral *TRPV6* as a genetic factor in racial cancer disparities due to excessive cytosolic Ca^{2+} when calcium consumption is high. A synthesis of recent discoveries in *TRPV6* structure/

> function/evolution and association of increasing calcium intake with a pattern of increase in high-risk prostate cancer in AA men supports her hypothesis and leads us to suggest a mechanism for it. Hilliard's hypothesis, if true, would imply a rethinking of current dietary advice delivered to people of African ancestry, and should be investigated via research into the impact of *TRPV6* haplotype on prostate, triple-negative breast and other carcinomas from which people of African ancestry suffer disproportionately. However, to our knowledge, no studies have been conducted to investigate the association of the *TRPV6* ancestral haplotype with cancer, whether as a main effect or in interaction with another gene or an environmental factor, such as calcium intake. If such research were to uncover an association of the ancestral *TRPV6* haplotype with cancer in people of African descent, this could lead to precision medicine inclusive of precision prevention delivered through improved cancer risk assessment and dietary recommendations specific to the ancestral genotype.[12]

During that same period, I scanned the periodically updated Genome-Wide Association Studies catalogues, which contained comprehensive listings of data analyses involving disease-causative gene triggers. The 2022 edition's entry for TNBC listed a sample number of 4,928 Europeans and a replication sample number and ancestry of 3,457 Europeans. However, no editions contained entries for the African *TRPV6*a calcium ion channel.

Several earlier studies published in journals found that carriers of the African variant not only absorb more of the calcium ion Ca^{2+} but expel less of that ion in the urine, retaining more in the skeletal structures. The renal retention of calcium appears to be caused by ethnic-specific variants on the *TRPV5* renal-associated gene.[13] Since

Blacks retain more calcium than Whites and expel less in urine, it may be possible that the *TRPV6* gene can become overloaded with these active free calcium ions.[14] So what happens when the African variant of the *TRV6* gene is exposed to more calcium ions than it can take in?

I thought about Mama. She loved sour cream and would eat a cup of it for breakfast and two or more cups during the day in addition to three cups of coffee enriched with cream. If the calcium in the nondairy foods she ate during the day added up to 200 mg., then her daily total calcium intake would have been 987 mg. (that is, 675 mg. [sour cream] + 28 mg. [coffee with cream] + 84 mg. [scoop of ice cream] + 200 mg. [nondairy foods])—just 13 mg. shy of the 1,000 mg. of calcium a day for females recommended by the US Department of Agriculture (USDA). But what if Mother's West African ancestral intake of the nutrient guided by a more highly absorbent African variant of the calcium ion channel did not require a daily calcium intake of 1,000 mg.?

Although we do not know my mother's exact ancestry percentages, her family consisted of fair-complexioned Louisianans of slave descent, some of whom had, over the generations, quietly "passed" for White. If she inherited 50 percent African/50 percent European DNA, her body's daily calcium needs could be closer to 600 mg. (that is, 100 mg. from African ancestry + 500 mg. European ancestry). In that case, her daily consumption of calcium would have been 1.6 times higher than her body's biological requirement. It is also possible that her physician, in noting that she did not drink milk, would have suggested that she increase her daily calcium intake to 1,200 mg.—the USDA-recommended level for postmenopausal females. Had that been the case, my mother would have consumed two times more calcium than her biological set point defined by ancestry. If calcium needs vary according to whether one carries the more calcium-absorbent African or the non-African *TRPV6* variant, then many Black women in the United States may have found themselves in a precarious situation—though at low risk of osteoporosis and other bone diseases, they have a high susceptibility to TNBC, which has not seen the same therapeutic breakthroughs as other types of breast cancer.

A 2018 study comparing *BRCA* (the BReast CAncer gene) and TNBC concluded that the *BRCA* mutation carrier status had no effect on survival rates in patients with TNBC, even though oncology protocols continued to encourage *BRCA* testing for all American females. The report also noted that *BRCA*-associated cancers had been found to be responsive to platinum-based chemotherapy. But such was not the case with TNBC.[15] In short, the most aggressive and fatal form of breast cancer—TNBC—for which Black women are at highest risk was not affected by breakthroughs being made in patients with *BRCA*-associated cancers.

MISSED CUES

We cannot know whether a calcium intake of either 160 percent or 200 percent above biological requirements as defined by admixed ancestry would trigger mutagenesis of the *TRPV6*a African gene variant in my mother's (or any other Black woman's) breasts. However, we can at least be sure of three facts.

First, as of 2022, neither the African *TRPV6*a calcium ion channel variant nor any other gene variants not found in Europeans are being investigated by American researchers as possible triggers in Black females' unusually high susceptibility to TNBC (although researchers in China are investigating the possible role of the *TRPV6* gene in cervical cancer and other advanced solid tumors).[16]

Second, high-profile advances made in the treatment of *BRCA*1 and *BRCA*2 breast cancer have had little if any impact on the mortality rate of Black women with TNBC (as genetic therapies are available for *BRCA* but not for TNBC, which continues to rely on the less efficacious surgery, chemotherapy, radiation therapy, and targeted drugs). In short, Black women diagnosed with TNBC will also have a *BRCA* variant. But the *BRCA* therapies used on women of European ancestry do not prolong the lives of TNBC patients.[17] Mathieu Lupien, a senior scientist at the Princess Margaret Cancer Centre in Toronto, remarked, "This disease [TNBC] has no precision medicine . . . So, patients are treated with chemotherapy because we don't have a defined therapeutic target. Initially, it works for some patients, but close to a quarter of patients recur within five years from diagnosis, and many develop chemotherapy-resistant tumors."[18]

Third, no one in America's medical research establishment has asked the possibly life-or-death question related to calcium: does an overconsumption of calcium (based on a genetic population's biological need rather than USDA-standardized dietary calcium-intake levels) contribute to *TRPV6*-expressing cancers in Americans who carry the African *TRPV6*a calcium ion channel variant? Why not even consider the possibility? One conceivable answer is found in an article entitled "*TRPV6* Gene Variants Do Not Influence Prostate Cancer Progression," which gave the impression that researchers investigated both the African (*TRPV6*a) and the non-African (*TRPV6*b) variants of the gene to come up with their findings.[19] However, a careful reading of the study's supplemental data revealed that no links to cancer were found in the non-African *TRPV6*b gene variants of its German subjects, *while the homozygous African* TRPV6*a variant was not tested because of its rarity in European populations.*

I counted a total of 5,601 journal articles published on Medline dealing with either African Americans/Breast Cancer (3,278), African Americans/Triple-Negative Breast Cancer (248), African Americans/Prostate Cancer (2323), or African Americans/Metastatic Prostate Cancer (59). None of the African American/Breast Cancer or TNBC articles examined the African *TRPV6*a variant or any other possible cancer triggers that might be present in Africans that were not already identified in Europeans. But there are a few rays of hope. A June 2021 commentary by medical researchers Loren Saulsberry and Olufunmilayo Olopade argued that a significant barrier to achieving health equity in cancer genomics and pharmacogenomics is lack of access "to truly personalized care and tacit acceptance of a precedent whereby health systems, equipped with inadequate research from diverse populations, can only react to, rather than avoid, suboptimal cancer outcomes for underrepresented patients."[20] There has been no formal response to their comments.

African American females may in fact be most deeply harmed by the rigid medical paradigm defined by the biological genotypes of Northern Europeans. It had taken decades for me to realize that the problem might revolve around the differing calcium needs of diverse genetic populations. But I remember all too well when I first became aware of what might turn out to be critical information regarding calcium's effect on an altogether different biological process.

PREGNANCY COMPLICATIONS AND CORRELATIONS

During my years in Dakar, Senegal, I occupied a one-room apartment nestled in a garden of date palms and trumpet vines behind the Musée de l'Institut Fondamental d'Afrique Noire (renamed Musée Théodore Monod in 2007). Each day I took a bus along the Corniche, the scenic shoreline road that led to the research institute where I worked on my doctoral dissertation. While I was waiting for the bus one morning at my usual stop in front of a café, a Senegalese woman whose pregnant belly bulged under a vibrantly patterned grand bubu sat beside me on the bench, where we waited longer than usual for the bus. The woman reached into her satchel and offered me a rock of clay. Munching on it, I recalled female relatives in the Louisiana bayou country passing around chunks of Argo starch. In exchange for her generosity, I went into the café and purchased a pint-sized bottle of milk for her and a bottle of mineral water for myself. When I returned to the bench and offered her the milk, she scrunched up her face in what I can only describe as horror. I had committed an egregious faux pas, but I didn't know what it was.

Over time I came to realize that 99 percent of adults in Senegal were lactose intolerant. Not even pregnant women drank milk or ate dairy foods. Mothers breastfed their babies. And with the notable exception of the ethnic Fulbe (agro-pastoralists who inhabited the northern portion of the country on the border with Mauritania), dairy did not figure in the diets of the population. If this woman had lived in the United States or saw a US-trained physician, her lactose intolerance might be noted and she may have been advised to supplement her diet with low-lactose or lactose-free milk. In the minds of doctors following the protocol in which they had been steeped, this pregnant woman should at the very least be supplementing her diet with calcium. However, the same African *TRPV6*a gene variant whose greater calcium absorbency allows West Africans to maintain healthy bones on 250 mg. of calcium a day is also found in the placenta. Women with the African *TRPV6*a gene variant would not need to supplement their diets with dietary calcium. But the effect that the more calcium-absorbent African *TRPV6*a gene variant might have on the healthcare of African American women—especially women who are pregnant—does not exist in the medical literature.

The field of obstetrics is influenced by some of the same unspoken attitudes found in other sectors of American society. For instance, the few studies on West African childbirth subjects contained in US journals share a remarkable commonality of tone. It is an unspoken assumption that medical practices on the African continent, whether by traditional healers or even by Western-trained doctors, are primitive and unsafe. Consequently, when studies are undertaken, they conform to Western norms and assumptions. This is certainly the case when looking at the levels of calcium in pregnant women.

A PubMed search of articles with keywords lactate, pregnant, West Africa, and complications revealed that between 2000 and 2020, nine out of ten articles focused on calcium deficiency in West African mothers. Because the standard in Western societies is 1,000–1,200 mg. of calcium a day for pregnant women, medical researchers get stuck on the notion of calcium deficiency. As the average daily calcium intake of pregnant West Africans is 250–400 mg., problems in childbirth will automatically be attributed in American medical literature to calcium deficiency. Other potential causes will be overlooked, even if the presumed rational explanation makes no sense. For instance, one study of pregnant women in Gambia, West Africa, who were diagnosed as having low calcium intakes notes, "Calcium supplements resulted in significantly lower bone mineral content, bone area, and BMD at the hip throughout 12-month lactation. The women also had greater decreases in bone mineral during lactation at the lumbar spine and distal radius and had biochemical changes consistent with greater bone mineral mobilization."[21]

Why does the American medical system believe low calcium intake and lactose intolerance are such a serious medical problem for all pregnant women? Lactose intolerance can be a serious disease in populations of Northern European ancestry. Genetic trade-offs may be required for the human body to benefit from the introduction of dairy, which can be a life-saving renewable protein source. Calcium is the most tightly regulated mineral in the biology of humans. And yet dairy foods inject four times more calcium into our digestive systems than would have been customary in the early human diet. Given our body's tight control of cellular calcium, might nondairy populations, especially those carrying the more calcium-absorbent *TRPV6*a variant, be maladapted to the 400 percent calcium boost caused by a dairy food culture?

In seeking solutions to our nation's high maternal mortality crisis, we might consider the possible inflexibility of the obstetric protocols used by American doctors as being an issue worthy of further reflection. A January 13, 2021, article in the *Washington Post* with the headline "Mortality Rate for Black Babies Is Cut Dramatically When Black Doctors Care for Them after Birth" quoted convincing statistics to make its case but offered no penetrating insights into why that should be.[22] As a result, it could reinforce readers' sense that racism and discrimination are the only factors killing Black mothers—as opposed to those issues being added to inflexible medical assumptions.

Obstetrical protocols are presumed to be applicable to women of all demographic backgrounds. But what if they are based on the biological needs of females from European cultures? Most Blacks are lactose intolerant, which is pathologized in America. Pregnant Black females fall so far below the minimal USDA standards for calcium intake that White doctors, as a matter of course, advise calcium supplementation. However, because Black doctors, like many of their patients, are lactose intolerant, without adverse effects, they do not have the same nutritional preoccupation. What professional platforms or forums do Black and other minority physicians now have to compare notes on certain presumably standard medical practices that they reject? The fact that America's medical establishment has produced so few researchers of color drastically reduces the number of voices with the authority to challenge standard medical procedures that might adversely affect patients from their own communities. If the contents of the aforementioned *Washington Post* article prove correct based on further research that Black babies die less frequently when cared for by Black physicians after birth, I would posit that minority physicians already know why Black mothers fare better under their care. But the medical establishment does not allow minority physicians to step outside of their rigidly defined practitioner's lane and speak truth to power.

I am not suggesting that medical protocols advise pregnant Black women who show signs of pregnancy crises to reduce their intakes of calcium and possibly sodium. Such decisions should be made by physicians. But to fail to see potential clues in the fact that African Americans emanate from a genetic ancestry that consumes 75 percent less calcium and nearly 95 percent less sodium than Whites is reckless. Medical researchers should be examining these and other possible instances when the biological needs of Black women might differ from those of their White counterparts. Currently, it just does not appear that most doctors are being encouraged to think outside of the box when confronted with a minority patient whose symptoms or therapeutic results do not fit the textbook cases. We know, for instance, that expectant Black women are more than three times more likely to die in pregnancy and during postpartum than White women, five times more likely to die from pregnancy-related cardiomyopathy and blood pressure disorders than White women, and more likely to experience physical and emotional racism-related stress that might exacerbate potential complications of pregnancy.[23] According to the National High Blood Pressure Education Program Working Group on High Blood Pressure in Pregnancy, African Americans have a high risk of the following hypertensive diseases during pregnancy:

1. chronic hypertension
2. preeclampsia-eclampsia
3. preeclampsia superimposed on chronic hypertension
4. gestational hypertension (transient hypertension of pregnancy or chronic hypertension identified in the latter half of pregnancy)[24]

So although a national organization communicates high health risks for African American expectant mothers, there still exist no accurate supporting data that would take into account their non-European biological traits. Comparative studies have been undertaken of mortality rates among African women, but they sometimes lack rigor. Making statistical comparisons with maternal mortality in rural societies without hospitals makes it all but impossible to ask the most relevant question: Should we in the United States be examining our one-size-fits-all obstetrical protocols?

We cannot explain away the high mortality rates of Black women during and after childbirth by relying solely on socioeconomic disparities they face—which include the stresses of racism and lack of access to healthcare. While wealth has never served as armor against bigotry, it has at the very least offered the advantages of better medical care. Yet there are too many stories of well-known, wealthy Black women with top-notch healthcare suffering from pregnancy-related issues. In 2017, tennis star Serena Williams had blood clots and nearly died after giving birth to a daughter. That same year, entertainer Beyoncé Knowles developed toxemia and preeclampsia, causing her blood pressure to reach dangerous levels. And on January 28, 2017, Dr. Shalon Irving, a thirty-six-year-old Black epidemiologist at the Centers for Disease Control and Prevention (CDC), died after giving birth due to complications associated with high blood pressure.

By 2021, the CDC was reporting that 2.6 times more Black American women were dying in childbirth compared with their White counterparts.[25] The leading causes of death include hypertensive and cardiovascular disorders and type 2 diabetes. An editorial in the *New York Times* that same year headlined "Easing the Dangers of Childbirth for Black Women" included staggering updated statistics showing that Black women were dying from childbirth-related causes at twelve times the rate of White women.[26]

It is clear that we need more studies on the calcium-absorbent African *TRPV6*a gene variant and on African American healthcare in general. Studies to determine whether carriers of the African variant either benefit or may be harmed by calcium supplementation therapy have not been conducted. While a study of preeclampsia in African American women conducted in 1993 observed "an abnormal intracellular free calcium [Ca^{2+}]i metabolism as early as the second trimester of pregnancy," there is no evidence that follow-up studies to clarify the findings were conducted.[27] Virtually all searches in the PubMed and Google Scholar databases using the keywords "preeclampsia" and "calcium" called for the supplementation of intracellular free calcium as a preventative. No research was conducted or even suggested that might posit whether African American women responded differently to this protocol—with one notable exception.

Born in 1948, Lawrence Malcolm Resnick began studying at Northwestern University at age fifteen. Six years later, he had completed a medical degree from that institution and gained prominence in the ensuing years for his work in the field of cardiology and hypertension at Weill-Cornell Medical Center. But it was after a seven-year stint at Wayne State University Medical Center in Detroit that Dr. Resnick made what would have been greeted as one of the most stunning discoveries in health disparities research had he not died prematurely of pancreatic cancer shortly thereafter. Resnick's clinical research on the pregnancy complications in African American women was published in a January 1999 article entitled "The Role of Dietary Calcium in Hypertension." His work provided the first real evidence in the medical literature that Black women responded differently than White women to dietary sodium and calcium during pregnancy. He observed, "This analysis suggests that although increasing oral calcium intake to achieve at least current nutritional standards is entirely appropriate, uniform recommendations for all hypertensives to further increase or decrease dietary calcium or salt may be inappropriate and will obscure those for whom these maneuvers are particularly relevant."[28]

A 2013 study that further clarified ethnic differences in salt sensitivity stated, "Despite the different methods of determining salt sensitivity, studies have consistently found approximately one-quarter to one-third of all normotensives, 50 percent of all hypertensives, and up to 75 percent of non-Hispanic black hypertensives to be salt-sensitive."[29]

Subsequent studies have referenced superimposed preeclampsia in which chronic high blood pressure worsens during pregnancy, triggering additional health complications. However, no studies suggested that the standard treatment for White females, which involved supplementation with 100 mg. of aspirin and 2,000 mg. of calcium, be altered in any way for Black women who might exhibit different metabolic responses to dietary calcium.[30]

LACTOSE SUPREMACY

During my freshman year in college, a required reading assignment for one of my natural sciences courses was Thomas Kuhn's 1962 book *The Structure of Scientific Revolutions*. The author's use of paradoxes and anomalies to signal the moment at which a particular paradigm or way of organizing knowledge no longer worked somehow penetrated to the deepest layer of my being, where it sat for nearly five decades. Several years ago, I recalled that slender volume by Kuhn as I reflected on what medical researchers were calling the "paradox of African American bone health."[31] Federal nutritional standards labeled most African Americans as both vitamin D and calcium deficient, pointing out that 75 percent of this community suffered from lactose intolerance (or lactase nonpersistence). But at the same time, Blacks had the lowest rate of osteoporosis and the highest BMD levels of any American ethnic group.[32]

While a paradigm in medical science offers a framework for organizing knowledge, theories, and research methods, it can also act as an impenetrable wall with nearly magical powers to keep out interlopers. For instance, in all medical textbooks published in the United States, lactose intolerance is labeled a pathology because it limits the intake of dietary calcium. Unfortunately, there has been no way to know if supplemental calcium was contributing to health problems in Blacks. This mineral is far too entwined in American food culture to elicit any questions whatsoever. All of the articles accepted in peer-reviewed medical journals and listed in PubMed (a federally run and funded database containing more than thirty-two million citations for biomedical literature) associate the topic of lactose intolerance/lactase nonpersistence with pathology. Given this research profile, there isn't a category for research or hypotheses that lactose intolerance/lactase nonpersistence might be a neutral trait in certain populations. Our culture has supported drinking milk, eating yogurt, and supplementing calcium for decades. Thus, our medical system has made calcium an inviolate issue in health.

A related issue not currently under discussion but which should be is the USDA's vigorous marketing of dairy products, especially to societies in Africa and Asia that are 99 percent lactose intolerant but whose rates of fragile bone diseases are lower than in the West.[33] Even though companies that produce dairy products may market low-lactose dairy products to those societies, there may be ethical issues in the marketing of a product that not only may not be needed but could have adverse effects on a non-European–defined population. The diets of people in non-Western nations often do not include the high amounts of calcium that Americans ingest.

PO' FOLKS' FOOD

My Angolan-American friend Matamba and his wife, Regina, upon returning from a trip to West Africa, showed me photos of Matamba's gray-haired aunts, uncles, and younger family members hiking up and down an overgrown trail. These were not peasant farmers leading lives of exhaustive physical labor. Rather, Matamba's relatives were *assimilados* (that is, educated Angolan professionals). One was a retired medical doctor and several had retired from government service. Their adult children were accountants, businesspeople, and schoolteachers. They were active adults living in the capital city of Luanda, which, given the nation's oil and diamond wealth, was referred to in tourist brochures as one of the most expensive cities in the world. These middle-class Africans were not debilitated by the same chronic diseases from which my African American friends and family members suffered. And yet, weren't Americans presumably enjoying the highest living standard in the world? Not being a statistician, I was uncertain as to how to go about comparing premature deaths on both sides of the Atlantic from diseases related to obesity, hypertension, strokes, type 2 diabetes, and heart disease. But I could investigate nutrition.

Bryant Terry, African American chef and curator of *Black Food: Stories, Art & Recipes from the African Diaspora*, showcases recipes with common West African grains like rice, millet, cassava, sorghum, and corn. After reading his book, I decided to see what would happen if I tried out a diet more similar to that of my ancestors.

I tumbled out of bed the morning after a vegan-ish dinner of Brussels sprout salad with homemade, salt-free mandarin orange and ginger dressing. That slight exertion in and of itself was not normal. For years my wake-up ritual had consisted of sluggishly reaching for my cell phone and trying to clear my head. The difference in my energy level and state of mind so intrigued me that I kept up the new diet for the next few months. Given exposure over the years to insights regarding my West African ancestors' adaptation to a low-sodium food culture, I had over the course of the past two decades struggled and mostly succeeded in reducing my salt intake to below 1,000 mg. a day. But the new dietary changes called for eliminating cheese from my diet, an excruciating hardship, and seeking out meat substitutes. The latter became problematic because these processed plant-based brands invariably contained several times more sodium than the meats they were attempting to replace. However, no amount of Gas-X, Beano, and pre-washing every assortment of beans on supermarket shelves quelled the gas attacks I experienced. Just as I was about to give up on the diet altogether, something surprising happened—I started losing weight. It was not much; maybe half a pound a week. But I had completely lost my cravings for cupcakes with butter cream frosting, bagels with cream cheese, and the plasticized, so-called pastries on the baked-goods shelf at the convenience store. For a few weeks, I kept a food journal and learned that I had a ravenous response to gluten that could only be described as an addiction. Once I began consuming gluten-free bread and other products, my craving for carbohydrates melted away—a pattern I'd seen before. European colonizers had learned in the 1500s that they could wrest the world's most-fertile continents from indigenous hunter–gatherer populations using an early form of biological warfare, which involved the transmission of contagious diseases for which only farming populations had natural immunities and the disempowerment of these communities using alcohol. So, I pondered whether West Africans, who consumed nongluten grains such as sorghum, millet, and rice might have a similar addictive/craving response to grains with gluten (which medical researchers should explore further). But my lived experience was that cutting out gluten did in fact tamp down my cravings for baked goods and other carbohydrates.

In 2002, Judith Carney published the meticulously researched book *Black Rice: The African Origins of Rice Cultivation in the Americas* to acclaim in history and other humanities circles. The book provided an explanation for the insistence of South Carolina and Georgia plantation owners to only purchase slaves at auction from the Senegambian region of West Africa (also known as the rice coast). Professor Carney corrects the mistaken belief that Europeans introduced rice to West Africa and then brought the knowledge of its cultivation to the Americas.[34] Enslaved persons/farmers transferred both the seed and cultivation technologies needed to establish rice in the New World. Slave owners' cooks showed Southern whites how to steam rather than boil the grain and infuse dishes with savory ingredients.

By the nineteenth century, rice had become one of the most profitable crops in the South. So it is not surprising that the historical imagery of Blacks cultivating rice would find itself immortalized by the image of a smiling Black gentleman presumably named "Uncle Ben" on rice boxes in supermarket shelves across the United States. (The rice boxes were finally rebranded in 2020 and the Black man's image removed in response to public protests surrounding the death of unarmed African American George Floyd at the hands of the police.) The Southern states became the economic engine of the burgeoning nation due to the exhaustive efforts of enslaved persons/farmers. Popular historiographer Greg Timmons wrote that "slavery was so profitable it sprouted more millionaires per capita in the Mississippi River valley than anywhere in the nation."[35]

By the late 1600s, the mortality rate of early Native American enslaved persons/farmers (who lacked natural immunities from diseases common to Europe and Africa such as chicken pox and measles) on American plantations was close to 90 percent. European indentured servants, exposed to the ardors of Southern plantation life (including tropical diseases against which they had no natural defenses), succumbed at a rate of 50 percent. In contrast, the death rate of enslaved persons/farmers from the interior of West Africa was below 10 percent (as they carried natural immunities to the same agricultural diseases as Europeans, had some level of inherited protection against tropical diseases that were common in the Southern states at that time, and, of even greater value, had farming skills). Since these communities had produced sufficient food to feed themselves and their families from nutrition-poor, high-erosion lands with thin top soil back in their homeland, they were able to feed their families while enslaved. Modern genetic techniques were required to reveal the astonishing sophistication and selectivity of the trans-Atlantic slave trade. African Americans of slave descent have a genomic profile that is more than 99 percent farming populations emanating from a range of West African countries, with Nigeria being most prominent. But an even more valuable insight that can be gained from these test results is the fact that these enslaved Africans were farmers. Less than 1 percent of Black Americans' ancestral DNA is traceable to nonfarming hunter-gatherers who also inhabited the southern regions of West Africa.[36] During the era of the slave trade, West Africans were not kidnapped and sold willy-nilly. There was a vibrant market for enslaved persons/farmers, but only if they came from agricultural rather than nomadic or hunter-gatherer communities.

Dietary researchers have often used "prosperity" as a shorthand way of describing instances in which a minority population's health declines as it becomes acculturated to American society. But it is not only the higher standard of living in the United States that drove up rates of obesity and type 2 diabetes among African Americans and other minorities. It is also the freshet of American staple foods and the infinite choice of foods that a population, group, or individual may not be able to efficiently digest (as it/they may not possess the necessary gene variants and microbiome to do so).

With the 1920s movement of Black Southerners to the North and the accompanying abrupt shift in food culture during the era of the Great Migration, African American health has steadily deteriorated. Historian and author Jennifer Wallach of the University of North Texas tracks the dietary transition from rice and corn to wheat gluten. She noted that the early European colonists brought to the United States a preference for light-colored leavened wheat bread, which was perceived as high-status, upper-class food compared with other grains. This culinary hierarchy became engrained in American food culture over time.[37] It should therefore not come as a surprise that the early twentieth-century Blacks who entered the middle class adopted such attitudes. Wallach also observed, "[Booker T.] Washington micromanaged the dining plan for students and teachers at Tuskegee, advocating for their right to consume beef and wheat."[38]

As the civil rights movement gained steam after World War II and job discrimination barriers slowly fell, African Americans moved away from a diet of rice and corn because many in that community considered these non-gluten grains "Po' folks' food" (owing to the painful association of rice with slave plantations and corn with general deprivation). It was during this era that a quiet switch took place. The name "soul food" began being used to describe what were in fact the meaty and heavily wheat-floured dishes of White Southerners rather than the vegetable-dominated dishes found in traditional Black households. Upwardly mobile Blacks came to embrace this high-fat, wheat-gluten-based food culture with a speed and fervor that should have raised red flags among food historians. The *New York Times* published an article with the headline "Is It Southern Food, or Soul Food?" The following comments by soul food cookbook author Todd Richards and Southern food writer Virginia Willis elucidate the essential difference between the two styles of cuisine:

> Richards: People get preoccupied with skin color. But it's really a question of poor and wealthy. Only poor people would eat neck bones and chitlins. Fresh meat like chicken and pork was always a luxury.

> Willis: Fresh vegetables weren't a luxury; everyone had a garden, and the growing season is long. Across the South, it was an agrarian culture, for better or for worse.[39]

By 1992 observers were noting that little difference could be found between the types of foods eaten by Whites and Blacks. A 1996 article noted, "In 1965, there were large differences among groups in dietary quality, with whites of high socioeconomic status eating the least healthful diet, as measured by the index, and blacks of low socioeconomic status the most healthful. By the 1989–1991 survey, the diets of all groups had improved and were relatively similar."[40]

Physicians rely on family medical histories as a way to narrow down the number of potential disorders as they diagnose a patient's illness—a cost-effective strategy given the hundreds of chronic diseases, viruses, inflammations, and traumatic injuries that may present with a similar list of symptoms. Medical histories can be of even greater value when researchers assess radical shifts in the intergenerational nutritional history of defined populations. The insights that they impart can, in fact, dramatically cut the costs of eliminating racial and ethnic health disparities.

A foundational principle of modern nutritional studies is the concept of "dietary diversity." It operates on the assumption that cereal-based, presumably monotonous diets found in the developing world's poorer households represent the most serious nutritional problem facing humankind. An article in the *Journal of Nutrition* elucidates this guiding precept of nutrition:

> Dietary diversity (DD) is universally recognized as a key component of healthy diets . . . The few studies that have validated DD against nutrient adequacy in developing countries confirm the well-documented positive association observed in developed countries. A consistent positive association

> between dietary diversity and child growth is also found in several countries. Evidence from a multi-country analysis suggests that household-level [dietary] diversity is strongly associated with household per capita income and energy availability, suggesting that DD could be a useful indicator of food security. The nutritional contribution of animal foods to nutrient adequacy is indisputable, but the independent role of animal foods relative to overall dietary quality for child growth and nutrition remains poorly understood.[41]

In September 2018, an advisory issued by the American Heart Association amended this by now well-established principle with the following clarification, "'Eat a variety of foods,' or dietary diversity, is a widely accepted recommendation to promote a healthy, nutritionally adequate diet and to reduce the risk of major chronic diseases. However, recent evidence from observational studies suggests that greater dietary diversity is associated with suboptimal eating patterns, that is, higher intakes of processed foods, refined grains, and sugar-sweetened beverages and lower intakes of minimally processed foods, such as fish, fruits, and vegetables, and may be associated with weight gain and obesity in adult populations."[42]

Nutritionists and medical professionals in the United States are now aware of the degree to which Americans are consuming too much of the wrong foods—namely, products that are ultra-processed. Because variety is an operating principle of US nutritional research and culture, the very term "staples" refers to what people in poor countries eat. But the reality is that living organisms most easily digest the nutrients that have been most readily available to their multigenerational ancestors.

Every five years the US Department of Health and Human Services (HHS) and the USDA update *The Dietary Guidelines for Americans*, the goal of which is to reduce the high rate of chronic disease among Americans by bringing to the public's attention the latest nutritional findings presented as dietary advice. One outcome of this emphasis on diet as a means of combating such disorders as hypertension, cardiovascular disease, and type 2 diabetes has been the development of the Dietary Approaches to Stop Hypertension (DASH) diet—a food plan that emphasizes increased consumption of fruits and vegetables and lower consumption of high fat and sodium in the Western diet. In November 2022, the *American Journal of Cardiology* published a study showing that the DASH diet improved chronic disease risk in the American population and that its effects varied by sex and self-identified race. Coauthor Stephen Jurashcek noted, "Our study suggests that the benefits associated with these diets may vary by sex and race. While a diet rich in fruits and vegetables produced reductions in risk for women and Black participants, the effect with the DASH diet was twice as large in women and four times as large in Black adults."[43]

EXPLORING "OMICS"

Beginning in the 1990s, the collective efforts of scientists around the globe to sequence the human genome led to what is today referred to as the revolution in "omics" (a Greek suffix inferring "totality"). Over the course of the next thirty years, this transdisciplinary network expanded from the study of genomics (an organism's genetic information) to that of metagenomics (sequencing the microbial environment), proteomics (proteins), transcriptomics (RNA transcripts), and more.

The explosion of research on the microbiome—the microbes that naturally live in our bodies—supports the idea that populations of ecological niches may share the same microbes and, therefore, have the same food culture and ease in digesting certain foods. The incredible array of microbial forms found in the human body—including fungi, bacteria, and viruses—numbers at least 3.3 million. Medical researchers believe a person's microbiome may influence their susceptibility to illness and disease. Aries Chavira and colleagues reported, "The human microbiota (the microscopic organisms that inhabit us) and microbiome (their genes) hold considerable potential for improving pharmacological practice. Recent advances in multi-'omics' techniques have dramatically improved our understanding of the constituents of the microbiome and their functions. The implications of this research for human health, including microbiome links to obesity, drug metabolism, neurological diseases, cancer, and many other health conditions, have sparked considerable interest in exploiting the microbiome for targeted therapeutics."[44]

Not surprisingly, gene variants are often population-specific, as is the microbiome, which has its own unique taxonomy. Research has thus far identified at least three enterotypes in Europeans. I am pleased to note that, unlike the lag time in recognizing population-specific gene variants, warnings have already appeared in the scientific literature regarding the lack of transferability of data to other defined populations. Vinod K. Gupta and colleagues published a 2017 article entitled "Geography, Ethnicity or Subsistence-Specific Variations in Human Microbiome Composition and Diversity" that cautioned, "Some of the geographical/racial variations in microbiome structure have been attributed to differences in host genetics and innate/adaptive immunity, while in many other cases, [to] cultural/behavioral features like diet, hygiene, parasitic load, environmental exposure etc. The ethnicity or population-specific variations in human microbiome composition, as reviewed in this report, question the universality of the microbiome-based therapeutic strategies and recommend for geographically tailored community-scale approaches to microbiome engineering."[45]

The vestiges of racism and prejudice observed in a nation founded on slavery are soul wounds that will not disappear overnight. But American society does not have to achieve perfection to improve the quality of medical research provided to its non-European minorities. As for the African American community, too many clues are being overlooked because the lived experience of Blacks cannot be factored into genetic research if members of the Black community are not trained to participate in hi-tech medical deliberations involving their own health.

4 THE ALGORITHM

APPLYING GENETIC ANCESTRY TO CASE STUDIES

THE ALGORITHM—A SET OF RULES to be followed in calculations or other problem-solving operations—has been around since antiquity. From Babylonian mathematics in ancient Mesopotamia to Alan Turing's model of numerical computation and the Turing machine, we as a species have been solving problems and calculating data using carefully defined, step-by-step procedures for millennia. Machine-learning systems are primed with ever-increasing amounts of data, which give them the capacity to make analytical decisions or judgments. But if the data being fed into the computers reflect racial or sexual biases, the information output will be flawed and errors will be compounded—sometimes exponentially. For instance, algorithmic identifications of law enforcement mug shots are limited by the precision of the data input into the system. With several times more data on the facial features of Whites than Blacks, misidentifications in facial recognition software have led to such errors as Facebook labeling a video of African American males "nonhuman primates" and, more seriously, the arrest of innocent African Americans.[1]

The Big Data revolution burst onto the information technology scene at the turn of the twenty-first century. For the first time, exponential amounts of data could be analyzed and stored, transforming fields as far removed from one another as the sciences, agriculture, music, and gambling. The capacity to generate hypotheses from massive amounts of data has become foundational to the emerging field of precision medicine, which aims to provide personalized treatments for individuals by considering genes, environments, and lifestyles. In healthcare, the consequences of misdiagnoses can be devastating. (In 2019, the *Washington Post* exposed the problems caused by a presumably race-blind metric that further highlighted the inequities associated with Blacks' access to healthcare.[2]) And yet, the accurate, well-thought-out use of algorithms can over time transform the quality of healthcare and the precision of diagnoses for *all* Americans. In fact, one of the most vociferous complaints regarding the unfair application of algorithms to American minorities might in fact be rooted in the racial allocation of organs for transplant.

RACE-CORRECTION ALGORITHMS

My Texas primary care physician (PCP) did not diagnose me with kidney disease because, unlike the doctor in Japan, he applied a race correction algorithm to my kidney function. Use of a race correction algorithm is one of the most controversial subjects in American medicine because the scientific community has not come to terms with the chaos created by the use of "race" in medicine. The issue, rather than being "race" versus "colorblindness," is the crucial need to see classification systems as valuable tools that demand precision. The only reason using the term "Black" sometimes works in American medicine is because 89.2 percent of the time it captures a single ecological niche population (ENP) (that is, African Americans of slave descent whose common ancestry from the interior of West Africa gives them a shared high risk of hypertension, which has been highlighted in health disparities research). Black Americans of slave descent are also an admixed population comprising West Africans, Northern Europeans, and Native Americans. It is important to note that although the 2.1 million recent immigrants from sub-Saharan Africa are too genetically diverse to be described by any of the so-called racial algorithms being used in the United States, the use of such algorithms is growing. And yet, a 2019 study showed that when heart-failure-management algorithms are used, Black and Latinx patients with heart disease are less likely to be admitted to cardiology service.[3] Kidney transplants involve an allocation system of which race is an integral part (as research shows that kidneys from Black donors have higher graft failure rates than those from other groups). In obstetrics, the use of a vaginal birth after cesarean (VBAC) algorithm tends to increase cesareans for Black women—who have high rates of childbirth mortality. In urology, the STONE score is supposed to predict the likelihood of kidney stones in patients who present with abdominal pain.[4] Yet I had a personal nightmare several years ago when an emergency room physician suggested I go home and take Tylenol for what turned out to be an excruciatingly painful kidney stone.

Alas, here's the deal in cases such as the aforementioned: medical research that identifies different outcomes in diverse demographic groups could be real and valuable but is currently sloppy. When biological differences are noted, the drawing of boundaries in terms of to whom such differences are relevant becomes crucial. But since using sociological race categories seems to sometimes work, they continue to be used long after precision medicine has demanded that they be jettisoned.

THE KIDNEY TRANSPLANT PROBLEM

A June 2021 article entitled "Jordan Crowley Would Be in Line for a Kidney—if He Were Deemed White Enough," described how Crowley, a fair-complexioned, wavy-haired young man, was denied a kidney transplant because race-based algorithms give different, tougher scores to Blacks than Whites.[5] Many Americans were both confused and outraged to learn for the first time that different scores were being used to measure the severity of kidney disease in African Americans and that these scores made it more difficult to obtain a transplant. But I knew the truth, and in that case, racialized medicine was not the problem.

The racial classification scheme used in the United States—even in the sciences—is a direct byproduct of slavery. Two hundred years ago, plantation owners sexually assaulted female slaves and would then classify the children born of such liaisons as property (assets) on their balance sheets (a labeling practice today employed by both Whites and Blacks in America). Even though the census now allows Americans to self-identify in terms of race, seldom will African Americans of slave ancestry identify as anything other than "Black" as a matter of pride/a way of taking ownership of a descriptor for their skin color that a mere fifty years ago would have been considered an insult.

Prior to the incident involving Crowley, healthcare providers used the Estimated Glomerular Filtration Rate (eGFR), a tool that measures the level of an individual's kidney function and determines the stage of kidney disease. Questions arose as to whether Blacks were being discriminated against in the allocation of kidney transplants since eligibility for patients was based on having the lowest scores, and the eGFR raised Blacks' scores above those of Whites. The principal means by which eGFR measured kidney function involved assessing levels of creatinine—a waste product formed by the breakdown or wear and tear of muscle cells—in the urine or in the blood. Healthy kidneys take creatinine out of the blood and dispose of it through urine. If the level of creatinine in the blood is high, the kidney is not doing its job efficiently. But intake needs vary, especially those of African Americans of slave descent (because their ancestors adapted genetically to survive in a specific ecological niche). So even though elevated creatinine levels signal the presence of kidney disease, physicians (and the algorithm) need to consider what constitutes the normal range in diverse defined populations when analyzing results. Diagnosis and transplant allocation should consider the Apolipoprotein L1 (*APOL1*) gene variants an individual carries in their personal genome.

THE COLORBLINDNESS TRAP

Any form of medicine based on the scientifically discredited taxonomy of "race" will not serve anyone's purposes. On the other hand, advocating "colorblindness" in the field of precision medicine and ethnopharmacogenomics is not in practice what its proponents believe it to be. Allow me to clarify. While the fight against racial discrimination in all facets of American society has certainly resulted in progress since the era of slavery, we still have a way to go. The battlefield must be reconfigured. Precision and genomic medicine requires more technical definitiveness on the part of medical professionals than did medicine of the past (wherein traditional medical practitioners strove for a certain universality, which resulted in claims of discrimination when minority patients were treated differently from their White counterparts). For this reason, "colorblindness" became the rallying cry for those intent on fixing a racially broken medical system. But in the context of twenty-first-century personalized medicine, "colorblindness" has not come to signify treating everyone equally. Rather, it treats minorities using the same genetic measurements and standards as those used for Whites even when it became clear that human genotypes differed—not by "race" but by more subtle ecological factors not necessarily identifiable in a person's racial identity. In this new environment, "colorblindness" resulted in Blacks being indistinguishable from patients of Northern European ancestry for medical treatment purposes so as to avoid claims of racial discrimination.

As you may recall from the Prologue to this book, the Japanese physician who misdiagnosed me was unaware of the differences in kidneys of various ethnicities or defined populations relative to those of Asians. Japanese laboratory procedures and testing forms did not have an "African American" box to check off. While standard serum creatinine values for Whites in the United States are less than 1.3 milligrams per deciliter, they are even lower for Japanese—0.8 milligrams per deciliter. But African Americans, at 1.91 milligrams per deciliter, exhibit healthy kidney function at creatinine levels that are higher than the cut-off point for Whites and more than twice that of the Japanese.

Black Americans of slave descent, unlike more recent immigrants from Africa, will have a substantially higher risk of kidney disease from sodium overconsumption because they have inherited the *APOL1* G1 and G2 variants (adaptive variants that may have arisen only in West Africans when they settled in the low-sodium ecological environment of the continent's interior). Or it might be the case that they emerged from the deeper reservoir of the comprehensive African parental genome itself. In any case, the Black–White algorithm, including current adjustments, will remain distorted and unfair unless all racial designations, including those that purport to be nonracial but that do not use DNA ratios of genetic ancestry, are also discarded. Many so-called White Southerners are either unaware of (or hide) the fact that they carry some degree of African DNA. An individual whose family may have "passed for White" or who simply did not know that an earlier ancestor had secretly crossed the color line would then be shocked to learn that a hypothetical genetic ancestry test showed that they carry 90 percent European/10 percent African DNA. A study by Katarzyna Bryc and colleagues exposed this fact, which has only now become known because of DNA ancestry testing, and reported, "We note a strong dependence on the amount of African ancestry, with individuals carrying less than 20 percent African ancestry identifying largely as European American, and those with greater than 50 percent reporting as African American."[6] These percentages offer a sneak peek into the secret world of Blacks who "passed" for White. Let's remember that before 1865, any person born to an enslaved mother, regardless of how White their features, was an enslaved person. For that reason, individuals with sufficient European phenotypes whose family history was unknown in the state to which they migrated crossed the color line "passed" for White.

The Kidney Transplant Crisis

Jordan Crowley had predominantly Northern European DNA. His grandmother was White and other family members were admixed (African and European). (There are now sophisticated DNA sequencing tools that provide anyone willing to pay $99 with a genetic report delineating the precise ratios of specific European ethnicities to African and/or other DNA ancestries in their family lineage.) Had the actual ratio of Jordan's European versus African ancestry been taken into account in the scoring of his transplant eligibility, he would have met the qualifications for the new kidney. Instead, the draconian, all-inclusive definition of Blackness was applied in his case (as it is in all cases involving patients of admixed African ancestry who self-identify as Black). But a person like Jordan who may be 75 percent European/25 percent West Africa shouldn't receive the same kidney transplant score as a person who is 100 percent West African (as they would today).

Over the years, the medical community has grappled with rectifying the sorts of issues around race/ancestry that surrounded Jordan Crowley's case. Scientists Michael Yudell, Dorothy Roberts, Rob DeSalle, and Sarah Tishkoff called on the US National Academies of Sciences, Engineering and Medicine to convene a panel of experts tasked with eliminating the use of race in genetic research.[7] But given the complexity of the issue, progress has been slow. Fortunately, the scientific community now has the tools needed to identify genetic ancestry—and they perform with a never-before-achieved level of precision. But what has been lacking until now is a larger vision of how to use these methodological tools to their greatest effect.

Let's remember that Crowley had been labeled Black because of his interracial background, even though his percentage of European DNA was considerably greater than his West African DNA. The failure to take into account the European portion of his ancestry, even though biologically it affected his kidney transplant score, made him ineligible for a kidney transplant. However, we need to create a more scientifically accurate and noncontroversial (in racial and political terms) way of calculating transplant scores for Americans of mixed-race ancestry, and to acknowledge individuals' differences.

For decades, medical researchers hypothesized that Blacks had higher levels of creatinine because of greater muscle mass (since the biochemical is a natural breakdown product of creatine phosphate in muscle tissue). However, we haven't studied enough of the African genomic profile to know which Africans, when stratified into genetic groupings rather than lumped into together, have greater muscle mass than non-Africans. Recent studies are now questioning whether the measuring of creatinine levels is in fact the most accurate means of determining kidney function in all humans.

A 2008 study conducted by the research team of Joy Hsu reported, "We hypothesized that adjusting for muscle mass and related factors would eliminate or reduce the racial differences in serum creatinine levels. . . . However, even after making that adjustment, Black patients had significantly higher creatinine levels." Hsu concluded, "The higher creatinine levels in black patients compared to non-black patients could not be entirely explained by differences in age, sex, body size, or muscle mass."[8] The issue came under even greater scrutiny in 2019 when the authors of an article in the *Journal of the American Medical Association* reiterated the need to come up with a more equitable formula that did not repeat the racial injustices of America's past.[9] But what might that formula be?

Task forces set up by the National Kidney Foundation and the American Society of Nephrology have been developing new guidelines. And in September 2021, the National Task Force at the University of Pennsylvania Perelman School of Medicine introduced what it is calling a "race-free calculator," which draws on data from twenty studies as a means of gauging accuracy. But is there a more scientifically precise approach to this matter?

Nonracial Scientific Methods of Delineation

Given the conundrum surrounding race, would it be possible to develop a tool that takes into account a patient's ancestry in the only way that truly counts—biologically? This would involve stratifying populations for diseases and other medical purposes according to shared gene variants that impact a particular condition. Genome-wide association studies work poorly on African Americans because the databases involved are calibrated to match gene variants found in Northern Europeans for the targeted disease symptoms. Could we replace the "one-drop" (legacy of slavery) rule with a mathematically precise way of defining admixed populations?

It appears that the most impartial and unbiased statistical method of allocating risk factors for Blacks and other ancestrally admixed individuals is the use of DNA ancestry ratios. The alternative is to continue relying on the antiquated one-drop rule—which unfairly denied Crowley a kidney transplant and asserts that anyone who self-identifies as "Black" will be treated with non-European medical protocols even if the individual's DNA ratios are 5 percent West African and 95 percent Northern European. (Medical practitioners would consider such an individual's risk of sickle cell anemia equal to that of someone with 100 percent West African ancestry.) A March 2019 study exposed the implications of this fallacy, observing, "Local adaptation to new environmental conditions—including pathogens—has generated strong patterns of differentiation at a specific loci, that is, fixed positions on a chromosome where a particular gene or genetic marker is located."[10]

More than twenty years ago, John Aloia at New York University Langone Health suggested that studies produced more accurate results when admixture marking in African Americans was used to separate their African and European heritages. But as no follow-up to Aloia's suggestion appeared in the medical literature, in 2017 I developed and tested a protocol that would standardize the process of differentiating the DNA inheritance of admixed individuals for use in my own research.

THE ENP MODEL: A NONRACIAL TOOL FOR DIFFERENTIATING POPULATIONS IN MEDICAL RESEARCH

I developed a tool—the Ecological Niche Populations Model (ENP Model)—to track disease causation based on the genetic adaptations of human populations to unique ecological zones across the globe; it aligns well with the framework devised by the National Academies in their comprehensive 2023 report.[11] (*Note*: the ENP Model is not a tool for researchers attempting to section off the world for the sheer convenience of avoiding the use of racial categories. Rather, it identifies the ancestral etiology of certain racial and ethnic health disparities in the United States by clustering disease-triggering gene variants that differ from one defined population to the next while remaining fluid [that is, it is not a function of biogeography]. It delineates humans according to the medical condition or health disparity under investigation and includes admixture ratios created by intermarriage.) For instance, West Africans inhabit the Tsetse Belt, which is unsuitable for cattle breeding or dairy farming. Typically, this population is 99 percent lactase nonpersistent (lactose intolerant). My research has identified certain cancers that might be triggered by the overconsumption of calcium in African Americans carrying the more highly absorbent *TRPV6*a calcium ion channel variant even though this group's consumption of the mineral is deemed deficient by US standards (see Chapter 3). Carriers of the *TRPV6*a variant descended from inhabitants of the West African Tsetse Belt would thus constitute a specific ENP—which might overlap with but would not be precisely the same ENP for investigators researching hypertension and kidney failure in African Americans whose sodium-retentive *APOL1* gene variants are a function of ancestral adaptations to the low-sodium interior of West Africa but not the salt-rich Atlantic coast.

The ENP Model can be used to identify patterns of disease susceptibility linked to gene variants in a population cluster that has adapted to, for instance, particular ancestral diets, certain pathogens, high altitude, high levels of arsenic, or other such things considered hazardous to most humans. It is for this reason that the West African/calcium ENP might overlap with the West African/sodium ENP. But while the former looks at the genomic profile of inhabitants of the nondairy-consuming Tsetse Belt region, the latter only draws boundaries around West Africans of the interior who adapted to a portion of the continent that did not produce halite (or salt). Even though the African American (of slave descent) ENP identifies a population whose ancestors emanated from West Africa, it differs from the West African/sodium ENP because African Americans (of slave descent) carry a unique average admixture profile (75 percent West African, 24 percent Northern European, and 1 percent Native American).

The ENP Model's algorithm identifies the particular high-risk health disparity that a certain community or group (but not amorphously defined race) might have in common. Thus, African Americans of slave descent can be aggregated for medical purposes as they share these historical, genomic, and medical traits: their ancestors were skilled agriculturalists rather than hunter-gatherers or members of nomadic tribes; they emanated from the sodium-deficient interior of West Africa (unlike recent immigrants to the United States from the capital cities of modern Africa, who would have emanated from the coast); they, as ancestral DNA testing will find, carry West African ancestry and some degree of admixed DNA (that is, mixed with Northern—as opposed to Eastern or Southern—European DNA); they may or may not possess some Native American DNA; and they may possess *APOL1* G1 and G2 gene variants that previous studies have identified as playing the primary causative role in raising hypertension and kidney failure rates in this American demographic group.[12] A large number of these individuals also carry the African variant of the *TRPV6*a calcium ion channel and numerous other variants not found in non-African populations.

Defining an ENP

The African American/Sodium-Metabolic (Disparities) ENP is defined by four shared components—the shared geographic factor; the admixture factor; the shared disease factor; and the shared disease-triggering genomic profile. The process of applying the ENP Model begins with data from racial/ethnic health disparities research—a vibrant new field in medicine that calls attention to health differences among American demographic groups. Many disparities are related to lack of access to healthcare and racial discrimination. The ENP Model offers a genomic determinant to be taken into consideration alongside the vital social determinants of health like poverty and racism—both of which are deeply entrenched in American society and not likely to be reversed in the short term. However, genomic triggers can be arrested in many cases either with nutritional changes or with medication. (See Figure 4.1.)

It is important to remember that the ENP Model is not about skin color or phenotypes but rather about shared disease risks. And the susceptibilities are linked to a shared genetic adaptation to a particular ancestral environment. For instance, an African American interested in calculating an appropriate calcium intake for him- or herself might consider admixture ratios as a function of the amount of dietary calcium consumed by each segment (300 mg. a day by West African ancestors, 1,200 mg. a day by European Ancestors, and 1,100 mg. a day by Native American ancestors). Rather than the USDA-recommended 1,000–1,200 mg. a day, the African American would consider an ancestrally adjusted level closer to 521 mg. a day. According

The Components of an
ECOLOGICAL NICHE POPULATION (ENP)

Defining an African American/Sodium-Metabolic Disparities ENP

FACTOR "A"	FACTOR "B"	FACTOR "C"	FACTOR "D"
Shared Geographic Factor 37 million descendants of enslaved persons/farmers brought to the United States before 1865 from low-sodium interior of West Africa	**The Admixture Factor** Median admixture ratio of: 74 percent West African 24 percent Northern European 1 percent Native American	**Shared Disease Factor** High rates of hypertension, strokes, and kidney disease	**Shared Disease-Triggering Genomic Profile** G1, G2 variants APOL 1 gene

Figure 4.1 **The ENP Model can be used to bypass unscientific racial definitions in studying diverse populations.**

to genetic mathematician Xuexia Wang of Florida International University, this calculation can also be viewed mathematically as:

$$C_{admixed} = \sum_{i=1}^{n} prop_i \times C_i$$

where $C_{admixed}$ is the general nutrient value or kidney transplant score for an individual of admixed genetic heritage; n is the total number of ancestries for the individual of admixed genetic heritage; $prop_i$ denotes the global ancestry proportion for the ith ancestry; and C_i denotes the identified bone-healthy calcium intake or kidney transplant score in the ith ancestry.

To test out this algorithm, three hypothetical case studies (calculating more accurate sodium intake, calcium intake, and kidney transplant scoring) in an individual of admixed ancestry for which the numerical values may differ from one defined population to the next are presented in Table 4.1.

Table 4.1. Applying the ENP Model: Three Case Studies

Case Study #1A: Connie's Adjusted Dietary Sodium Intake Based on Ancestral Ratios	
Step 1	Genetic ancestry data show Connie, who self-identifies as an African American, as having the following ancestry: - 25 percent West African (11 percent Yoruba, 5 percent Malinke, 6 percent Wolof, 2 percent Hausa, and 1 percent Ovimbundu) - 75 percent Northern European (12 percent Irish, 35 percent English, 10 percent Danish, and 18 percent German)
Step 2	Research data show that the standard sodium intake for an average West African farmer from the interior of the continent during the seventeenth and eighteenth centuries was 200 mg. a day. Further research shows that the typical sodium intake of Northern Europeans has on average been 3,400 mg. a day.

Table 4.1. *(continued)*

Step 3	(.75 × 200) + (.25 × 300) = 1000 mg. of sodium a day. This value represents a healthier daily sodium intake for Connie than either the current US average of 3,400 mg. a day or the USDA-recommended <2,300 mg. a day.
Case Study #1B: Connie's friend Lani's Adjusted Dietary Sodium Intake Based on Ancestral Ratios	
Step 1	Genetic ancestry data show Lani, who self-identifies as an African American, as having the following ancestry: - 10 percent West African, 80 percent Northern European, and 10 percent Native American.
Step 2	Research data show that the standard sodium intake for a typical West African farmer from the interior of the continent during the seventeenth and eighteenth centuries was 200 mg. a day. Further research shows that the typical sodium intake of Northern Europeans has on average been 3,400 mg. a day. Anthropological data on traditional coastal Native American sodium intake is estimated to be 2,500 mg. a day.
Step 3	(.10 × 200) + (.80 × 3400) + (.10 × 2500) = 2900 mg. of sodium a day (this value represents a daily sodium intake for this admixed individual that will be closer to the USDA-recommended <2,300 mg. a day).
Case Study #2: Dietary Calcium Intake Based on Ancestral Ratios	
Step 1	Genetic ancestry data show that Connie has 75 percent West African and 25 percent Northern European DNA.
Step 2	Research data show that the standard calcium intake for a healthy/typical inhabitant of the West African Tsetse Belt during the seventeenth and eighteenth centuries was 250 mg. a day. Further research shows that the healthy/typical calcium intake of Northern Europeans has on average been ≥1,000 mg. a day.
Step 3	(.75 × 250) + (.25 × ≥ 1,000) = 437.5 mg. of calcium a day (this value represents a healthier daily calcium intake for this admixed individual than the current USDA-recommended intake of 1,000–1,200 mg. a day).
Case Study #3A: Connie's Kidney Transplant Eligibility Score	
Step 1	Connie, who self-identifies as African American, has an admixed ancestry that is 25 percent West African and 75 percent Northern European. Connie applies for a kidney transplant.

(continued)

Table 4.1. *(continued)*

Step 2	The United Network for Organ Sharing (UNOS) is the private, nonprofit organization that makes decisions on organ transplants. It stipulates that an individual would need to have an eGFR score lower than 20 to be placed on the organ waitlist. The UNOS uses one proprietary algorithm for Blacks and another for Whites to equalize kidney transplant opportunities among these American demographic groups as Blacks suffer kidney failure at a higher rate than Whites. Connie's eligibility score is calculated to be 21.
Step 3	Connie is declared *ineligible for transplant.* Given the same level of kidney function, had Connie been considered White and the corresponding algorithm been used, her eligibility score would have been 17 and she would have been eligible for transplant. But use of the algorithm for Whites is not an option because of the one-drop-rule/a percentage of her ancestry is African.
Case Study #3B: Connie's Kidney Transplant Eligibility Score (using ENP Model)	
Step 1	A genetic ancestry lab test confirms that Connie has an admixed ancestry that is 25 percent West African and 75 percent Northern European. Connie applies for a kidney transplant.
Step 2	Connie's ancestry is more accurately defined by the UNOS as 75 percent European and 25 percent West African. Her eligibility score for the portion of her ancestry that is European would be .75 × 17 and her eligibility score for the portion of her ancestry that is West African would be .25 × 21. Therefore, as a person of European/West African admixed heritage, her Black and White eligibility scores will be proportioned and calculated as follows: .75 (17) + .25 (21) = 18.
Step 3	Connie is declared *eligible for transplant.*

To use the AA ratios, a new algorithm would be required to determine West African to European DNA ratios in African Americans of slave descent. The current use of a "Black" score is too imprecise. That is, an African American whose ancestors came to the United States on slave ships can hypothetically have anywhere from 99 percent European DNA and 1 percent West African ancestry to 100 percent West African ancestry taking into consideration the one-drop rule.

Caveats

I offer two caveats for users of the ENP Model:

1. The data required to identify the average sodium, calcium, or other nutrient intake of an ENP or defined population are not always available (even though the World Health Organization compiled nutritional data by country). I could use the model on West African and African American populations *only* because, as an African specialist, I have personal experience as well as access to knowledge of both populations from written texts and anthropological sources as to the average sodium intake in different segments of West Africa from the sixteenth century onward. But in many other geographical regions around the world, the appropriate data on nutrition and other environmental factors has yet to be compiled. Thus, it is vital to recognize that the aggregation of effective data for ENP models represents a transdisciplinary undertaking.
2. Given the fickle nature of inheritance, there is no scientific basis for assuming that the offspring of a Northern European blonde and a dark-haired West African will have medium-brown hair, medium-brown skin, and medium-brown eyes. This rule of partial randomness also applies to myriad other genotypes, including nutritional needs, kidney function, and other biological values that may differ among defined populations. For this reason, the ENP Model and its corresponding algorithm are *not* to be used casually or as ancestrally based food and lifestyle tips for healthy Americans. Who knows what particular gene variants a healthy individual of admixed heritage may or may not have inherited? Nevertheless,

> the ENP Model represents a powerful diagnostic tool for medical practitioners treating individuals who display a certain profile of symptoms or syndromes whose etiology may bear ethnic clues.

Despite the aforementioned caveats, this methodology will nonetheless prove to be invaluably beneficial in decreasing health disparities within our multigenomic nation. This is because the ENP Model removes the fallacy of universality upon which American medicine, including modern genomics and precision medicine, is based. Non-White estimates of adaptive biological values have been practically nonexistent because, until now, the US medical community did not see the need for data about ecological conditions in countries to which immigrant populations might be adapted.

Caveats aside, the use of an updated algorithm or calculation such as the ENP Model that takes ancestral genomic data into consideration is imperative for the health of all Americans. Just as medical practitioners would never imagine basing a diagnosis on a patient's self-reported blood pressure levels, the practices of precision and genomic medicine and ethnopharmacology must raise their own standards of performance to avoid making important diagnostic and therapeutic decisions based on an individual's self-identified race or ethnicity. Certain populations might even be suffering from health issues caused by the fact that they are following medical and nutrition advice that is not in line with their biological needs. (For example, had I remained in Japan, whose population consumes on average three times more sodium than White Americans and 50 times more of the mineral than my West African ancestors, the Japanese creatinine tests could have eventually landed me on a dialysis machine even though I possess two healthy kidneys.)

The ENP Model, while not perfect, will undoubtedly improve on self-identified racial classifications as new datasets describing non-White populations in greater detail become available. But the removal of "race" from medicine should not occur at the cost of devaluing the racial, ethnic, and cultural identities of Americans who have good reason to feel pride in their ancestry.

5 ANCESTRAL GENOMICS

EMBRACING A MULTIGENOMIC NATION

TWENTY YEARS AGO, the scientific world was unaware of the sheer range of genetic variation in our species. However, as the field of genomics has grown, so has the realization that much of the accumulated data collected in the Genome-Wide Association Studies comes from Americans of European descent, who have an advantage over Asians, Africans, Native Americans, and other groups. That is, only four Y-chromosome DNA haplogroups account for most of Europe's patrilineal descent. More specifically, virtually all Americans of European descent belong to the first haplogroup (R1b), found primarily among Western Europeans (also referred to as Northern Europeans).[1] That data cannot be universalized nor applied fully to populations from other ancestral backgrounds. And yet we know that personalization to the level of the individual has become one of the most critical factors determining the quality of healthcare in this century. Even so, medical practitioners in the United States should not be faulted for having failed to master every detail of African history and ecology that might prove relevant to the health of some members of my community. But the medical community should be accountable for approaching human health in the hi-tech, data-driven twenty-first century as though Western (Northern) Europeans, one group among hundreds of human groupings around the globe, represent the prototype of all human biology.

A change in perspective is needed, given our knowledge that the same diseases in humans are triggered by different gene variants in diverse defined populations. And to make matters more complex, these divisions do not correspond to racial divisions. Our deeper understanding of the human genome opens up new opportunities to advance health. But we must first grapple with the provocative counter-intuitiveness of the medical and health tasks that confront us. The reason is not malice (as innovative gene-editing techniques now show promising results with sickle cell anemia) but rather the fact that we as Americans have not given ourselves permission to put together the pieces of any biological puzzles that do not frame White bodies. Consequently, the medical literature relating to minorities is data-poor and riddled with unexamined oddities, paradoxes, and anomalies. Once the data stop making sense, the tendency among researchers has been merely to call for ever more research based on the same ill-fitting hypotheses and paradigms.

The ten steps outlined below are aimed at closing the disparities gap in medical care and advancing healthcare for all. They require commitment but are paradigm-changing, achievable, and, surprisingly, low in cost.

REPLACE RACE IN HEALTH DISPARITIES RESEARCH WITH ECOLOGICAL NICHE POPULATIONS (ENPS) OR OTHER RELEVANT TAXONOMIES

I lack the expertise to write a book about the ways in which minute variations in the genetic architectures of the Beijing Han, Northern Han, and Southern Han might be responsible for the variety of therapeutic responses to what appear to be the same disease. Having devoted my entire professional life to exploring the evolutionary history of West Africa and African Americans of slave descent, I can say with certainty that medical researchers in the United States need to recognize the value of applying historical and ecological precision to our nation's demographic palette. When we draw the wrong boundaries around population groups, we lose clues that will lead to identifying crucial patterns of disease susceptibility.

The demographic mosaic of American society enriches the quality of all our lives. Therefore, we should not need to apologize for affixing descriptive labels to our ancestral backgrounds, ethno-linguistic communities, or sexual orientations because this diversity deserves to be celebrated. In cases of historical trauma, we must speak aloud what needs to be fixed. The use of racial and ethnic labels serves a valuable purpose politically, culturally, and socioeconomically. But genetics and medical research are different kettles of fish. We should learn to appreciate the new technological tools that allow us to understand at the deepest level of precision which nutrients best nourish us, which medications have the highest levels of efficacy given our gene variant triggers, and which therapies take into account the entirety of our ancestral DNA. Absent this, race becomes an impediment to accomplishing goals in the more scientific domains. It is imperative that medical practitioners understand the distinctions among ENPs—people who would otherwise be nonscientifically classified and misdiagnosed as belonging to the same so-called race. The ENP Model is neither a disguise for race nor strictly geographical; it acknowledges the intricate dance among three distinct factors that may or may not overlap from one case to the next: geographical patterns, admixture ratios created by intermarriage, and groupings defined by shared gene triggers for specific diseases. In this era of artificial intelligence, such computations are easily achievable so long as we are developing the needed non-European databases.

Biologist Joseph Graves is one of the nation's foremost advocates of training medical students in the basic principles of evolutionary biology so they can understand the sometimes-nuanced but nonetheless critical factors involved in the use of racial concepts in both the practice of medicine and society at large. Graves explains, "While we would expect there to be alleles [gene variants] of medical relevance that are differentiated by ancestral geographical factors, we would not expect there to be drugs with impacts that are race-specific as has been frequently claimed. This does not mean that there are not differences in how individuals or populations respond to drugs; it simply means that these responses do not correspond to socially defined notions of race." He goes on to say, "A particularly troubling example of this ongoing misconception was revealed by the fact that in a recent study one-half of the medical students surveyed harbored false beliefs concerning biological differences between socially defined racial groups. In conjunction with these false beliefs they rated the pain of 'black' lower than the pain of 'white' patients and as a result made inappropriate treatment recommendations for the 'black' patients."[2]

A bizarre and possibly even life-threatening example of racialized medicine involved the formal training received by X-ray technicians throughout much of the twentieth century. According to Otto Juettner in a 1906 textbook, "the skin of the negro offers more resistance to the X-rays than non-pigmented cuticle."[3] In the 1940s, dental students were taught to increase X-ray exposure times for Black dental patients because "their oral tissues were more resistant to x-rays."[4] In the 1950s and 1960s, X-ray technicians were instructed to use higher radiation levels to penetrate black bodies. Even the 1961 and 1963 editions of the General Electric Company instructional manual advised the use of increased radiation on Black patients.[5]

Retiring the use of race in the genetics and genomics fields is not meant to dismantle all human distinctions, as doing so would merely bring us back to the default setting—the DNA profile of Northern European Americans. Instead, we must expand our knowledge of ancestral genomics and build our databases accordingly.

An article in the *New England Journal of Medicine* included the following observations regarding the common but unfounded belief that ignoring race would reduce health disparities: "We urge our colleagues in medicine and science to refrain from haphazardly removing race from clinical algorithms and treatment guidelines in response to recent initiatives attempting to combat anti-Black racism. The ultimate goal is to replace race with genetic ancestry in an evidence-based manner. But we have not yet reached a point where genetic-ancestry data are readily available in routine care or where clinicians know what to do with these data."[6]

STRATIFY RATHER THAN DIVERSIFY DEFINED POPULATIONS

In common usage, "diversifying" means adding ethnic or racial representation to an otherwise homogeneous referent, which works for integrating, for example, classrooms, workplaces, and neighborhoods. However, for scientists in the field of human genetics, diversity counts the number of gene variants within a given ethnicity or community in relationship to the aggregate number possessed by its species. The consensus among scientists is that the greatest degree of diversity, measured in numbers of polymorphisms (gene variants), is found among the inhabitants of Africa. (The numbers of polymorphisms decline as a measurable gradient defined by increasing distance from the African continent, which supports the Out-of-Africa theory of our species' origins.[7])

The discovery that Africans carry the largest amount of genetic diversity is so paradigm-shattering for some Americans that, when discussed, the subject is all too often obfuscated in technical jargon. For instance, the literal meaning of African American would include recent immigrants from coastal West Africa, the Swahili coast of East

Africa, Egypt, Somalia, Madagascar, and the Tuareg nomadic tribes of the Sahara. However, it is simply not possible to seriously study the hypertension crisis in so-called African Americans of slave descent without drilling down to the discrete demographic population that exhibits these symptoms at the risk level targeted for research intervention.

We must identify the differences in genetic variants using an algorithm that defines both the geographic boundaries of the population under discussion and the particular high-risk health disparity that they have in common. We would then be able to stratify the defined populations using genetic tools to identify the specific disease or symptom-causing gene variants they share with one another but not with others (even though the others may share their outward racial phenotypes).

IMPOSE STRICTER QUALITY CONTROLS

Because the one-size-fits-all paradigm in modern medicine remains unacknowledged, it is seldom evaluated for accuracy. Even though racial/ethnic health disparities research has become a serious endeavor in the priorities of medical institutions, too many studies relating to Blacks and other minorities are repetitious. For instance, dozens of studies have appeared in the medical literature over the course of the past fifteen years announcing that Blacks and Whites do not share the same gene-variant triggers for such-and-such a disease. If we took the time to look, we would even find that recent immigrants from coastal West Africa exhibit different disease-triggering variants than African Americans of slave descent whose ancestors emanated primarily from the interior of the continent. Too much time and research funding are being devoted to noting Black exceptions to the flawed one-size-fits-all paradigm. *JAMA Neurology* published an article identifying novel DNA regions for Alzheimer's disease in African Americans found in non-Hispanic Whites.[8] There is no indication that this discovery was followed up even though older Blacks suffer twice the rates of dementia as Whites. In fact, July 2023 headlines across the United States announced new groundbreaking treatments for Alzheimer's. But the clinical trials upon which the advances were based had screened out Black patients with higher rates of the disease. The reason given was that they exhibited different responses to beta amyloid, which represented a defining characteristic of the disease in Whites.[9]

Recent studies have also found different disease-triggering variants for NAFLD (nonalcoholic fatty liver disease), cardiovascular disease, hypertension, breast cancer, prostate cancer, gestational diabetes, type 2 diabetes, strokes, and so on.[10] Too little attention is given to pursuing a solution-oriented approach rather than a let's-do-more-research-on-the-same-unscientifically-classifed-race-using-the-same-failed-research-design approach when the stated goal is to find therapeutic solutions to the medical problems of America's minorities.

Another stumbling block that may be hindering breakthroughs in medical research regarding African Americans despite the Human Genome Project's many advances is methodological. One of the most elemental principles of statistical analysis is that correlation does not imply causation. The fact that two traits occur together does not tell us that one necessarily causes the other. Both may be influenced by the same outside factor as well as other possibilities. However, this fallacy in logic appears to occur far more often in research attempting to identify the etiology of certain genetic diseases or symptoms in Blacks than in research conducted on Whites.

As stated, the aggregation of data sets on diseases and gene-variant triggers is largely focused on a single defined population—Americans of European descent. African Americans and Asians will have different variants causing what appear to be the same diseases compared with Whites. Blacks will also have up to 50 percent more variants than Europeans, virtually none of which have been studied, sequenced, or even identified. What does this mean in practical terms? The National Human Genome Research Institute (NHGRI), which maintains a diversity-monitoring website, clarified the real diversity dilemma in the following way, "Sequencing of Khoi-San bushmen showed that even two people from adjacent villages were as different from one another as they were from any two European or non-African ancestry individuals."[11]

A 2012 article presented research findings that linked the use of chemical hair relaxers by Black women to uterine fibroids.[12] The research data were compelling but for one detail that I quickly pointed out in a brief letter to the journal's editor. I was not altogether surprised when, several years later, I read in an article in another journal that Black women's use of chemical hair relaxers caused ovarian cancer. Once again, I pointed out the fallacy—correlation does not automatically mean causation. The statistical correlations between this disease and the use of hair relaxers appear persuasive, but merely on the surface. If we look more closely, design flaws emerge that would only be noticeable among those with a more intimate familiarity with the culture in question. For one thing, differentiating the study subjects into "Blacks" and "Whites" injects concerning levels of imprecision into the data. In such cases, the use of ancestral DNA ratios instead might eliminate much of the confusion. In any case, the individuals categorized as "White" will characteristically have close to 100 percent European DNA. However, those categorized as "Black" will range from having 90 percent European ancestry and 10 percent African ancestry to having 0 percent European ancestry and 100 percent African ancestry. Designing the study in such a way will only cause confusion if the research aim is to determine certain causal relationships. To test the hypothesis that chemical hair relaxers increase the risk of certain cancers in African American females, it is crucial that we reckon with certain cultural factors. For instance, White women might use chemical hair relaxers to smooth out wavy hair. But that is seldom the case with African American women, who generally find wavy hair, though not 4c hair/hair that has the tightest curl pattern, aesthetically appealing. The more European DNA a self-identified Black woman possesses, the wavier will be her hair in most cases. Let's remember that the point of this exercise is to find the cause of rates of ovarian cancer in Black women that are higher than those of White women. Statistical analyses will only show that Black women with the highest percentages of European DNA do not use hair relaxers and have a lower incidence of ovarian cancer. So what has been proved? Well, common sense tells us that rubbing a chemical compound on any part of our bodies is not what nature intended. But our understandable fears of fibroid tumors and uterine and breast cancer can make Black women vulnerable to casually designed, costly taxpayer-financed medical studies that move us no closer to a solution to the health issue confronting us. Now that the National Institutes of Health has reissued previous studies linking relaxers to various cancers in Black women, enterprising law firms have begun targeting African American female cancer sufferers in advertisements for class-action lawsuits against the manufacturers of chemical hair relaxers.[13] (While the law firms may benefit from such actions, it is unclear whether Black women will learn that the real causes of their cancers are being overlooked or misascribed.)

It is simply not clear what institutional body in the United States speaks up for or even evaluates the quality of medical research in genetic medicine targeting minority communities. I am, of course, referring to cases where certain medical needs of minorities diverge from those of the mainstream. This has not always been true, given the impressive breakthroughs made with more-easily-diagnosed sickle cell anemia.[14] However, in twenty-first-century medicine, it appears that the more political pressure Black politicians place on the medical establishment to attend to the needs of constituents, the more funding is made available for vague, sociological studies rather than research into new and innovative hypotheses that lead to successful therapies.

Medical practitioners are human and, as is the case with all of us, sometimes make mistakes. But it becomes crucial to point out errors that fall into a recurring pattern when they emanate from the imposition of faulty racial thinking. The careers of several promising Black athletes were dashed when they received erroneous diagnoses of a fatal heart condition known as hypertrophic cardiomyopathy. A 2016 article that identified five DNA variants common in Blacks that were mistakenly linked to the disease stated that although there are many benefits from genetic testing, there is also a risk that variants may be misclassified. The authors acknowledged that multiple patients, all with African or unspecified ancestry, received positive reports as a result of the misclassification and noted, "The misclassification of benign variants as pathogenic that we found in our study shows the need for sequencing the genomes of diverse populations, both in asymptomatic controls and the tested patient population. These results expand on current guidelines, which recommend the use of ancestry matched controls to interpret variants. As additional populations of different ancestry backgrounds are sequenced, we expect variant reclassifications to increase, particularly for ancestry groups that have historically been less well studied."[15]

As for research focused on reducing health disparities in the African American community, a common but often fruitless approach is combing all of sub-Saharan Africa for answers—as such an approach represents the "racial classification fallacy" in medical genetics. Because the African genome is heterogeneous and encompasses many times more gene variants than may be found in Europeans, it should not be assumed that the populations of Lagos, Nigeria, Accra, Ghana, or even Nairobi, Kenya, will offer more clues about health disparities found in African Americans of slave descent. In most cases, they will leave us with even more questions unless we have identified a well-defined ENP (which, even here, will not encompass a so-called race). Our identification of ENPs should be attached to specific health disparities. If the disparity is the high rates of hypertension and kidney failure in African Americans, most inhabitants of the African continent, with the exception of the readily definable ancestors of Black Americans, do not dwell in low-sodium environments. Most East and Southern Africans do not inhabit malarial areas; they lack natural immunities to those pathogens and do not carry the sickle cell gene. Africans outside the Tsetse Zone of West Africa will neither exhibit immunities to trypanosomiasis nor necessarily be lactase nonpersistent (as would those within that region). Looking in Kenya for gene-variant triggers for triple-negative breast cancer in African Americans will be productive only if the hypothesis upon which research is based relates to correlative factors other than race.

PRIORITIZE ETHNO-PHARMACOGENOMICS

We all have the potential to react well or poorly to a particular drug intervention. Given the surprising degree of malleability found in human DNA, it is also true that different groups of humans, each having adapted to their own unique ecological environments, may evolve different gene variants to perform the same or similar functions. It is for this reason that population-specific differences arise in drug metabolism, resulting in a wide range of responses to medications (noting that certain medications can be more or less effective and may result in fewer side effects for some groups than for others).

The ability to note patterns that might encompass an entire defined population or ecological niche rather than a single individual is invaluable as it simplifies a process that might otherwise prove overwhelming—the juggling of DNA variables and drug responsiveness data points receding from billions to several thousand or fewer. But this macroprocessing of certain medical knowledge works only if we define our taxonomies with fluidity. That is, the boundaries must be determined by commonality of specific disease-triggering gene variants rather than skin color, other physical traits, or continent of origin. The scientific community now has sophisticated tools that can distinguish a West African from the salt-poor interior from his coastal neighbors. The medical research field needs to make use of this more granular knowledge.

Jonathan Kahn's *Race in a Bottle: The Story of BiDil and Racialized Medicine in a Post-Genomic Age* ignited a fire that is still not under control. Kahn asserted that profit motivated pharmaceutical companies to artificially segment the drug marketplace for ethnic groups, including in instances where a drug might be found to be less effective than originally promised for the larger American market.[16] In such cases, political pressures or inducements could be used to market such products in minority communities despite the lack of evidence that the communities would benefit from them. Such machinations on the part of corporate giants, according to critics, might include expensive advertising with unsubstantiated claims targeting the smaller, more vulnerable market. Kahn's book amounted to a devastating attack on "racialized medicine" and a celebration of "colorblindness." His message gained a substantial following in the medical community, particularly among progressive medical groups.

Two years later, African American cardiologist Clyde W. Yancy reluctantly stepped into the fray by publicly endorsing isosorbide dinitrate/hydralazine (BiDil), an antihypertensive and heart medication. It was the first drug approved by the US Food and Drug Administration to treat self-identified Blacks suffering from congestive heart failure. Endorsements from a Who's Who of African American organizations—including the Congressional Black Caucus, the Association of Black Cardiologists, the National Medical Association, and the National

Association for the Advancement of Colored People—soon followed. However, critics claimed that BiDil represented the repugnant concept of racialized medicine, wherein African Americans would receive inferior diagnoses and treatment. While Yancy's response to critics was that the drug worked, it was withdrawn from the market in 2009 because of poor sales given the reluctance of physicians to prescribe the controversial drug.[17]

The BiDil debate rages on to this day, with the drug representing the only medication for heart failure specifically indicated for African Americans. (It is important to note that Kahn and Yancy are experts in their respective fields who share the same goal of closing the disparities gap in Black healthcare.) And the search for new opportunities to market products by pharmaceutical companies might in some cases lead them to dump otherwise worthless drugs on vulnerable populations. The anti-BiDil group bristled at what they perceived as a dangerous trend—the commoditization of race in pharmaceuticals. But the Association of Black Cardiologists echoed Yancy's point of view, which was that the drug appeared to be effective for their patients. How to resolve the debate?

We must look at the issue within its larger context. Ethnopharmacogenomics happens to be one of the most promising therapeutic arenas for closing the racial/ethnic health disparities gap. BiDil may or may not be what its promoters claim it is. But the most efficient way to find out is to dig more deeply into evolutionary biology, jettison notions of race, and identify the specific gene variants or other factors that trigger divergent responses in diverse ENPs.

In 2014, members of the Eighth Joint National Committee issued a tentative set of guidelines suggesting that patients self-identifying as Black should not be given Angiotensin-converting enzyme (ACE) inhibitors as a first-line treatment for high blood pressure and heart ailments. Physicians observed that Blacks tended to be less responsive than Whites to ACE inhibitors and had potentially fatal swelling under the skin (that is, angioedema). However, eight years later, research conducted by Alice Adeles at the University of California San Francisco refuted the idea.[18]

Here was further proof that the issue of genetic differences in drug responses was gaining momentum. In previous generations, critics pointed to the exploitation of minorities in medical research (referring to use of their bodies in unethical experiments). But times have changed. Black and Brown bodies are no longer of value in medical research. Medicine based on universal traits in humans has been superseded by medicine performed at such a high level of precision that the data provided by the wrong defined population would be as worthless as using Listerine to cure throat cancer.

Modern pharmacology needs to be treated with a level of sophistication commensurate with growing scientific knowledge of human biological complexity. Medical practitioners have prematurely discontinued providing certain drugs to Blacks on the possibly ill-informed presupposition that they might be harmful to them. Such may currently be the case with azathioprine—a commonly used immunosuppressive medication whose known side effects are low white blood cell neutrophils counts, which increases vulnerability to infections.[19] But the problem that has arisen is the possibility that certain African populations naturally maintain lower neutrophils counts than do Europeans. As precision pharmacology gains speed, problems such as this will inevitably crop up; they can only be solved by instituting protocols that delineate human groups according to causative-gene-variant clustering, which would involve the use of the ENP Model or other nonracial taxonomies.

DE-EUROPEANIZE NUTRITIONAL GUIDELINES

Before the advent of refrigeration, the immediate ancestors of many defined populations—including Europeans and coastal Africans—consumed an average of 3,000–5,000 mg. of sodium a day and used salt both as a preservative and as a primary ingredient in meat processing. Scandinavians, Japanese, and certain oceanic islanders consumed upward of 10,000 mg. of sodium a day. But what if a specified population inherited traits that allowed their ancestors to farm hardscrabble land in the sweltering heat of the tropics on a dietary sodium intake of 200 mg. a day? What if these same people now lived in an environment where the average daily sodium intake is 3,600 to 4,100 mg. for men and 2,700 to 3,200 mg. for women? The result would be sodium toxicity, and the health implications for the population would be chronic and include high rates of hypertension, kidney failure, and cardiovascular disease—which are now seen in African Americans of slave descent.

What would it cost America's nutritional experts and the federal government to communicate that a community suffering from astronomically high levels of salt-sensitive hypertension and kidney failure that consumes several times more sodium than members of its healthy West African ancestral community might be dying prematurely of sodium toxicity and related diseases? Very little. No public health or medical research entity in the United States has to date assumed such responsibility. The most common reference in the medical literature to salt-sensitive hypertension in African Americans is the call for more research. Although footnotes occasionally found in highly technical medical journals suggest that Black Americans' sodium intake is lower than that of other members of our society, research in the same publications asserts that salt restriction increases risk of death from heart failure and type 2 diabetes and may raise LDL (bad) cholesterol and triglycerides.[20] Clearly, more research needs to be done. And as one medical directive does not work for all, the risks associated with certain medications and behaviors need to be communicated to individuals—and their physicians—so that they can begin to participate in their own healthcare.

In October 2021, the US Department of Agriculture and food manufacturers agreed to a slight reduction of sodium in processed foods to help the public meet its goal of reducing daily sodium intake from 3,400 mg. to 3,000 mg.[21] While the resultant 400 mg. reduction will offer genuine health benefits to many Americans—Whites, Latinos, Asians, Native Americans, and even recent immigrants from coastal African countries—it will be statistically unnoticeable to Blacks of slave descent whose overall health has reached a crisis stage.

USE ECOLOGICAL CLUES TO AVOID LOGICAL FALLACIES OF CAUSATION

African Americans have only marginally benefited from new discoveries in twenty-first-century genomic medicine because the most dramatic medical breakthroughs are hiding in plain sight. For example, a 2013 article hypothesized that salt sensitivity in relation to sodium absorption might be occurring in African Americans.[22] But that finding was quickly overshadowed by a more eye-catching hypothesis that appeared the same year in an article that linked the function of the African variants to trypanosomiasis (or African sleeping sickness) resistance (despite the fact that the 2013 article linked the *APOL1* variants to precisely the same hypertensive kidney disease symptoms that I had noted in African Americans emanating from the low-sodium interior of Africa).[23] The medical community chose to prioritize the linking of these gene variants solely to trypanosomiasis, thus thwarting any therapeutic potential related to following the variant's sodium metabolic function that might impact ailments common in Blacks (that is, hypertension, cardiovascular disease, and end-stage renal failure). And yet the biological truth of the situation could not so easily be glossed over. The ancestors of Black Americans survived in the sweltering heat of the low-sodium interior on a daily sodium-intake level deemed by US medical textbooks too low to sustain human life. Because geneticists were still using racial designations, they had no way of differentiating coastal Africans, who shared similar sodium-intake patterns with Europeans (3,400–5,000 mg. a day), with interior West Africans who adapted to a daily sodium-intake level of 200 mg. By failing to recognize the unique ecological niche occupied by the ancestors of African Americans, researchers never asked how the health of African Americans might be affected by the fact that their salt intake was 1,700 percent higher than their ancestors'.

In fact, scientists had noted certain variants that resonated with my historical knowledge of the genetic adaptations that the ancestors of American Blacks made to their ecological environment. But these facts were shunted aside as minor and possibly irrelevant observations. Why? Because the geneticists did not have access to the larger picture and were struggling to identify isolated pieces of data rather than recognize the possible discovery of crucial missing pieces in our understanding of health disparities in the African American community.

In 2013, several leading scientists suggested that laboratory tests for vitamin D resulted in African Americans being diagnosed as vitamin D-deficient. An article in the *New England Journal of Medicine* reported, "Low levels of total 25-hydroxyvitamin D are common among black Americans. Vitamin D-binding protein has not been considered in the assessment of vitamin D deficiency."[24] The article explained that American laboratories measure vitamin D levels indirectly by recording the blood levels of 25-hydroxyvitamin. The levels of 25-hydroxyvitamin D become a proxy for the availability of the nutrient in a person's body and are used because of the greater complexity of directly measuring amounts of vitamin D-binding protein. However, a more direct method of measurement would show that the actual usable vitamin D levels in Blacks are normal and only appear to be otherwise when the shortcut that works for Whites, but not necessarily other genetic populations, is used.

Black Americans like myself are still being prescribed vitamin D by primary care physicians because of normalized laboratory practices. Questions have arisen regarding the effects of deficiency of this particular nutrient in Blacks as they do not tend to suffer from the fragile bone diseases characterized by vitamin D deficiency. Some members of the medical community have suggested that, in the case of Blacks, vitamin D deficiency is likely responsible for the population's high susceptibility to hypertension, type 2 diabetes, and certain cancers. The diseases for which Blacks are at highest risk are known, and funding for research aimed at closing the racial/ethnic health disparities gap is becoming more readily available. However, the hypotheses upon which such research is based tend to mix and match the latest genotype found in Blacks but not Whites to an item on the Black susceptibility list and presume, without evidence, a causal link. African Americans are not being ignored in this new era of precision medicine, but its impact on their health remains minimal.

INVESTIGATE POSSIBLE LINKAGES BETWEEN DIETARY CALCIUM AND *TRPV6*-EXPRESSING CANCERS

America's one-size-fits-all medical paradigm is not designed to question what lies outside its boundaries—such as the linkage between African Americans' substantially higher rates of *TRPV6*-expressing cancers (including metastatic prostate, triple-negative breast, colorectal, and ovarian) and the uncontrolled proliferation of the protein expressed by the African *TRPV6*a variant.

Why should this hypothesis be given consideration? First, the medical community currently offers no specialized therapeutic interventions for these aggressive, *TRPV6*-expressing cancers. Second, the medical literature offers no other hypotheses to explain the calcium paradox in African Americans. Yet we know six relevant facts that support the linkage between dietary calcium and *TRPV6-expressing* cancers:[25]

1. The African variant of the *TRPV6* gene absorbs 25 percent more dietary calcium than the non-African variant;
2. African Americans, 75 percent of whom are lactose intolerant on average, consume 400 percent more dietary calcium a day than their (99 percent lactose intolerant) West African Tsetse Belt ancestors;
3. African Americans are considered, according to US federal nutritional standards, calcium deficient;
4. Calcium is one of the most tightly regulated substances in human cells;
5. Black Americans suffer a three to four times higher mortality rate than other US demographic groups from *TRPV6*-expressing metastatic prostate cancer, triple-negative breast cancer, colorectal cancer, and ovarian cancer; and
6. The ancestors of Black Americans, who on average have a daily dietary calcium intake of 250 mg., have high bone mineral density levels and minimal rates of osteoporosis.[26]

Combining these facts with what we currently know from the medical literature creates a foothold for a hypothesis linking overconsumption of dietary calcium with the *TRPV6*-expressing cancers. But unless the African variant of the *TRPV6* calcium ion channel is investigated and appropriate dietary calcium levels stratified according to the biological needs of diverse defined populations (as detailed in the previous steps), such a hypothesis may never come to light. However, if the African *TRPV6*a variant is the culprit in Blacks' higher susceptibilities to certain cancers, a promising therapeutic pathway might be testing the use of calcium channel-blocking drugs. (Because calcium is integral to apoptosis [cell death], use of calcium channel blockers [CCBs]/inhibitors was until recently counter-indicated in the treatment of all forms of cancer. Despite such concerns, there have been a few investigations of the potential links between calcium levels and certain cancers.)

A 2004 study by Jose D. Debes and colleagues found that in the case of prostate cancer, the daily use of calcium channel blockers reduced the risk by 40 percent.[27] A subsequent 2014 study observed that the action of CCBs went further by inducing cytotoxicity in androgen receptor positive cell lines) and concluded that this "may offer an innovative strategy for the treatment of fatal prostate cancer."[28]

The earliest effort to focus on inhibitors of the *TRPV6* gene's cancer-triggering action was made by Canadian researcher Jack Stewart, who in 2006 began testing a compound derived from the venom of the northern short-tail shrew.[29] A January 2020 article by Stewart offered a description of *TRPV6* as an onco-channel and stated, "Elevated *TRPV6* and subsequent sustained increases in cytosolic calcium activates the nuclear factor of activated T-cells (NFAT) transcription factors in cell lines of prostate 96 and breast cancers 93. In these studies reduction of *TRPV6* expression with silencing RNA reduced proliferation and increased apoptosis. Over-expression of *TRPV6*, which is constitutively active, results in a sustained elevation of intracellular calcium, which is required for activation of the calmodulin/calcineurin/NFAT pathway."[30]

In February 2021, the company founded by Stewart—Soricimed BioPharma—announced the successful completion of a phase 1B clinical trial of SOR-C13 (the shrew-venom compound) as a *TRPV6* blocker and cancer drug. That statement was soon followed by announcements by researchers at Chongqing Medical University in

China of successes using the local anesthetic lidocaine as a means of inhibiting the proliferation of *TRPV6*-expressing cancer cells.[31]

Stewart's research also identified the types of cancers that could be most affected by the blocking of the *TRPV6* channel, which include prostate, breast, pancreatic, ovarian, and colon.[32] Other possible *TRPV6* inhibitors are found in the medical literature, including the drug SOR-C13 (13 amino acid peptide), for which Stewart is the lead researcher in clinical trials of lidocaine, capsaicin, and ruthenium red.

However, outside of the investigations by Stewart and others, the medical research community continues to overlook the potential of *TRPV6* blockers to control the specific cancers to which Blacks are highly susceptible. As of 2022, these malignancies remain highly aggressive and without effective therapeutic interventions.

In short, the United States must find more efficient ways of directing pharmaceutical innovations along paths that benefit Americans from *all* ancestral backgrounds. As David E. Winickoff and Osagie K. Obasogie of the University of California Center for Genetics and Society warned, "We aim to promote public health and the societal interest in approving race-specific indications, but only when they are used cautiously and are supported by robust scientific evidence. Pharmaceutical science and biomedicine most certainly should not be colorblind. But they also must not be 'color-struck.'" They then added, "The battles waged to afford racial minorities 'equal protection of the laws' in the United States can be useful guideposts for regulators when considering race-specific drug applications."[33]

EQUALIZE SCIENCE EDUCATION IN AMERICAN SCHOOLS

Nearly 30 percent of White Americans adhere to some form of dogma that negates science and higher education. The statistics for Blacks and other minorities are quite similar.[34] And yet no one in American society deduces from such statistics that Whites should be deprived of a high-quality science education for "cultural" reasons. A 2020 article in the *International Journal of STEM Education* reported, "Although there has been a nationwide call for more diversity in the STEM [Science, Technology, Engineering, and Mathematics] fields for the past two decades, the results of these efforts have been slow and, in some cases, insignificant."[35]

Education, beyond allowing us to counter misinformation and understand our health and that of those around the world, is critical for future opportunities and aspirations; it's empowering. Science and medical education are on the highest rungs of self-actualization for any ethnic group in modern times because such knowledge ensures their survival. Long-term solutions to the dearth of minorities in the medical science fields must begin with nurturing and supporting students in inner-city school districts and those on Indian reservations and in poor rural communities, particularly as regards strengthening science, technology, engineering, and math education in their schools. This cannot be accomplished and supported merely through local efforts. It ultimately requires the commitment of our nation as a whole, beginning with the federal government and national medical, scientific, and educational establishments. Far too many Americans still cling to the notion that we can compete globally in science and medicine by importing our STEM talent from Asia and elsewhere. But that is the psychological legacy of a nation willing to raid underdeveloped nations for the specialized labor it refuses to invest in here at home. A 2015 study noted that "racial disparities in science achievement test scores begin as early as third grade."[36] Another study completed the same year stated, "These test score disparities were attributed to both socioeconomic status gaps between races and school qualities. In particular, Black and Hispanic students are more than twice as likely to live in low-income neighborhoods compared to White students, which directly contributes to less money for local public schools and indirectly less funding for STEM programs."[37]

In fall 2020, I organized a recorded workshop attended by a representative of the National Science Foundation. The meeting detailed special opportunities for high school students involved in STEM programs. I emailed the recording to schools that might have been overlooked in favor of those in more affluent districts. After four months, I finally gave up. While this task was insignificant in the larger scheme of science education, it broke my heart. My life's work as an African American professor had taken me around the globe, but feelings of guilt overwhelmed me as a Black mother. I feared that, over the years, a sense of urgency had been lost in my own commitment to working toward educational improvement for all young people. The telephone in the office of the principal of the first inner-city high school to which I'd sent the recording had been disconnected. The website of another had been taken down by the Internet provider, which replaced it with a message stating that the site could be reinstated once the bill was paid in full. Unlike schools in suburban districts, many of the inner-city schools had no social media presence whatsoever. Students in those schools were not being prepared to function in the computer-literate twenty-first century. Were my heightened sensibilities as a historian wrong in noting that the most destructive legacy of slavery is the system of mass incarceration that disempowers many young Black males, prematurely stunts their families, and, most importantly, results in children growing up without healthy male role models? A shocking 38.5 percent of young Black males are imprisoned at a cost to taxpayers of $33,274 per incarcerated person per year—a mere $3,000 less than in-state tuition for medical school.[38] Our legal system has not been able to rectify the fact that young Black males are incarcerated for crimes for which Whites are not prosecuted.[39] The funds that should be spent on STEM education in poor school districts are instead used to finance a demographically lopsided prison system.

The inadequacies of our educational system have compelled the medical community to become dependent on recruiting top physicians from developing countries (who, despite the numerous barriers they must overcome, constitute 40 percent of the American rural physician workforce).[40] The recruitment of medical professionals from Nigeria, India, and the Caribbean enriches our healthcare system on many levels. (The quality of my own healthcare has benefited over the years from having primary care physicians from different parts of the world.) Western medical training is in subtle ways enhanced by those from cultures that view the human body as a gift to be cared for with gratitude. Nonetheless, the current medical system should not be seen as a substitute for failing to dismantle American education's social ranking system, which makes it nearly impossible for inner-city Black Americans, Latinos, Native Americans, and rural Whites to aspire to become medical professionals.

IGNORE THE EUGENICISTS

No matter how vigilant our medical and scientific institutions, there will always exist an energetic, well-financed cadre that supports the eugenic distortions of human genetics. Why? A substantial portion, but hopefully not a majority, of the American electorate has a great deal of disdain for our multigenomic democracy. It favors authoritarianism in government and strict controls placed on scientific progress. Eugenics is a long-suppurating wound on our nation's soul whose roots trace back to the compromises the "democratic" founders of our republic made during slavery.

The scientific sophistries of those who insist on racializing humans and setting up hierarchies with themselves at the top will never heal the fears and insecurities emanating from their unpreparedness to embrace a democratic, twenty-first-century America. The scientific community should neither hesitate nor flinch when called upon to uphold ethical standards of data sharing. Academic freedom morphs into intellectual pornography when it is used to mislead a gullible segment of the American public into believing that they are justified in dehumanizing and victimizing the minorities in their midst. The National Institutes of Health's (NIH's) Database of Genotypes and Phenotypes (dbGaP) became a prime target for abuse by eugenicists because it has blocked the use of databases by researchers still bent on struggling to prove that high intelligence is a function of belonging to their self-identified race. Scientists should not have to waste time on the emotional bruises of sometimes well-educated but intellectually narrow-minded humans.

SUPPORT AFRICAN GENOMIC INITIATIVES

The scientific community now concurs that the San and Rainforest Hunter-Gatherer (RHG) groups carry the greatest amount of genetic diversity of all human populations (as they are the living descendants of the oldest surviving branch of early humans).[41] In the past decade, new research projects aimed at exploring further the implications of African genetic diversity for humanity as a whole have been initiated. The Human Heredity and Health in Africa (H3Africa) Consortium was formed in 2012 to address inequalities in global health and genetic research. The African Genome Variation Project, completed in 2014, added the sequenced genome of 92 individuals from 44 indigenous African populations to the reference genome. (Since then, it has expanded its databases and now includes 1,481 individuals and whole-genome sequences from 320 individuals across sub-Saharan Africa.) In 2010, the 1000 Genomes Project, the goal of which was to develop a comprehensive overview of human genetic variation that could be used throughout the biological sciences, including such disciplines as genetics, medicine, pharmacology, biochemistry, and bioinformatics, was launched.[42] The project reported producing 100 trillion base pairs of short-read sequence from more than 2,600 samples in 26 populations over a period of 5 years.[43] Another project that has expanded the genomic repertoire under investigation is the International HapMap Project, launched in 2002. It has created a database of 1 million genetic variations from the Nigerian Yoruba, Whites living in Utah, a Northern and Western European database compiled in 1980, the Beijing Han Chinese in Beijing, and Japanese in Tokyo.[44] In 2015, the NIH launched the All of Us project, which promised to recruit 1 million people across the United States to build what it referred to as one of "the most diverse" health databases in history. In 2018, the Human Pangenome Reference Sequence Project was initiated with the supplemental goal of producing high-quality sequences of approximately 350 genomes that would contribute to the creation of what has been referred to as a Pan-Human Genome.

While the number of African projects and their goals are impressive, they are also in some ways misleading. The funds available for all of these initiatives in totality are but a small fraction of the billions of dollars allocated to genetic research in the United States. And more tellingly, the Genome-Wide Association Studies (GWAS) Monitor was launched in 2020 to track diversity by recording the participation of diverse populations in these genomic studies in real time. As of 2023, individuals with European ancestry constitute 95.08 percent of total GWAS participants while those with African ancestry constitute only 0.17 percent.

Nigerian geneticist Charles N. Rotimi has taken a leadership role at the NIH to improve its diversity track record. Inducted into the National Academy of Science in 2018, Rotimi, who directs the Trans-National Institutes of Health (Trans-NIH) center for research in genomics and global health, was instrumental in founding the African Society of Human Genetics in 2003 and the Human Heredity and Health in Africa (H3Africa). Although his work has made inroads into resolving health disparities and addressing issues of genetic diversity, the path remains a rocky one.

The unusually low statistics for Africa in the GWAS Monitor show that our parental genome is still not being given the priority it deserves. We cannot merely count the number of new initiatives aimed at addressing diversity concerns. The stumbling block remains belief in the centrality of the European genome. When geneticists speak of inhabitants of Africa carrying the most comprehensive range of gene variants, they are ripping to pieces the conceptual foundations of eugenics. But when we learn that only 3 percent of that genome has been studied or sequenced, we see that America is still clinging to the essentialist notions of race upon which modern eugenics feeds. The irrational belief that our species' parental genome holds fewer immunological revelations than its out-of-Africa subsets will continue to haunt the field of genetics until the true nature of human genetic diversity is understood by all.

We must also consider the shortcomings in African genomic research caused by the fact that, as they are funded by Westerners, the agendas and the questions asked are being defined by non-Africans.[45] In 2014, a court ruling shook the African scientific community by drawing attention to dramatic instances of pay inequity when a Kenyan judge ruled against a United Kingdom–African research institute for maintaining a salary scale that compensated British scientists at a significantly higher rate than their African counterparts who were engaged in the same work.[46]

Given the African genomic roots of our species, far too little attention is devoted to mentoring African scientists and supporting intra-African collaborations.[47] Humans originated on the African continent, which retains more genetic diversity than any other continent. Millions of uncaptured variants have accumulated over the eons of evolutionary history that produced the human species. It should now be clear that the detailed study of African genomic variation is a scientific imperative.[48]

EPILOGUE

IN THE SPRING OF 1973, I arrived in London eager to scour the University of London's School of Oriental and African Studies' African collections for untapped sources on Timbuktu. Upon returning to my studio flat after long days of archival work, I was often greeted by the intoxicating aroma of *feijoada*. My culinarily savvy roommate, Jorge Sangumba, an Angolan medical student at the University of London, and I would attack the savory Portuguese stew of chicken, chorizo sausages, vegetables, and parsley while talking American politics. Jorge's opinions about my country's past and future were so far removed from my own that it hardly seemed like we were referring to the same place.

My childhood summers spent in the Jim Crow South and the undergraduate college years of pushing back against traditionalists had left me battered and hardened. Jorge, who had studied in America and traveled around the United States enough to have experienced white supremacy, segregation, and filthy, dilapidated bathrooms with "Colored" signs tacked to the door, seemed to have come away from that exposure with a passion for America. One evening after we had put down our forks and silently savored our stew, he turned to me and asked, "How old is democracy in America?"

"Two hundred years," I replied.

"You see, that is where you are wrong. I would calculate that democracy in your country is closer to eight years old," Jorge said. My raised eyebrow prompted him to continue.

"Americans mean well. But they exaggerate in ways that sometimes obscure the answers to their own problems. Your country established the rule of law for all its citizens in 1965. Is that not the year the Voting Rights Act was passed?" I reflected on the question for a moment, then nodded.

"Democracy is in its toddling stage and de-tribalizing can be so very painful. Do not be so harsh with this fragile, young child," he said. I pondered Jorge's meaning long after civil war in Angola had ended both his life and our fledgling marriage. It took time to heal from my losses and even more decades of life experience to fathom Jorge's words. I finally came to see that what he had always celebrated in America was aspirational rather than real, but beautiful nonetheless.

When Americans speak of races, they are referring to a vague classification scheme consisting of Europeans, Asians, Africans, and the Latino populations of South America. In the seventeenth and eighteenth centuries, when European immigrants first arrived on American shores, their racial self-identifications consisted of English, Irish, Prussian, Dutch, French, Polish, Ukrainian, and so on. Before being kidnapped and enslaved, West Africans self-identified as belonging to races like the Ashanti, Malinke, Hausa, Igbo, Fon, Yoruba, Wolof, Kimbundu, Ovimbundu, and others. These diverse ethnolinguistic groups inhabited different geographical locales within West Africa and often perceived of one another as being at least as foreign as the Irish would have perceived the Spanish. However, the rigid demands of the trans-Atlantic slave trade distilled the races of West Africa into one—Blacks—just as it had more passively done with America's European nationalities, who self-identified as White.

Jorge was speaking of the process that prompted Americans to de-tribalize. Despite the many setbacks, that process has in recent years been recorded in such books as *How the Irish Became White*, *How Jews Became White and What That Says About America*, and *The Wages of Whiteness: Race and the Making of the American Working Class.*[1] The politics of Europe in general and the Ukraine/Russian war in particular have magnified the tribal loyalties that delineate European ethnolinguistic groups. Decades ago, controversial affirmative action programs had allowed students (like me) to get a Harvard education. Our nation's path remains an agonizing one. But if we look around the globe and then back in history, we will see the truth. It is only in the multicultural aspirations of Americans committed to democracy that we find the number of human tribes shrinking rather than reinforcing their territorial and ethnolinguistic boundaries. That is progress.

TRUTH AND RECONCILIATION

In the eighteenth century, the United States was a young nation struggling to build a form of government that rejected monarchies and royal blood lines as Americans denied that its economy was built on the kidnapping and enslavement of West African farmers. Black Americans were told nothing about our ancestral homeland. Teaching an enslaved person/farmer to read was an offense punishable by imprisonment. Even as recently as the 1950s and 1960s, US textbooks did not include the history of the African continent or the peoples' ancestors. Africa was expunged from historical studies.

But there was progress, even as history repeated itself in shorter, more dizzying cycles. Harvard University filled a gaping hole in its curricula in 1971 by establishing one of the first departments of African and African American Studies in the United States, thanks to pressure from minority students of my generation and younger members of the faculty. But fifty years later, this field of study is now under attack as it has been reframed in parts of the United States as a dangerous concept—critical race theory.

If we as a country revert to the pretense that thirty-seven million Americans of African descent had no history, we do a grave disservice both to our past and to our collective future. The history of human populations and their African parental genome can lead us to knowledge that can be used to combat some of the most lethal disease susceptibilities found both in the African American community and within human biology at large.

A nation's medical paradigm is one of the most enduring symbols of its values. America's one-size-fits-all medical model does not yet benefit all Americans equally. Despite the challenges that lie ahead, America may yet fulfill its promise.

In June 2022, I returned to Cambridge, Massachusetts, for my fiftieth college reunion (which had been delayed a year due to the COVID pandemic). I strolled across Harvard Yard, exited through the wrought-iron gates onto Quincy Street, and found myself in front of the impressive Barker Center (the second floor of which houses Harvard's Department of African and African American Studies). I recalled having been an eighteen-year-old student fighting those who stigmatized the diversity and inclusion policies that had allowed me and others like me to set foot on the campus. I recalled the collective struggle of fellow Black students who fought to fill the gaping hole in history textbooks that had dismissed us as a people without a story. But instead of rekindling my youthful ambivalence toward the university, I was grateful for the lessons my time there had given me and had an appreciation for the sacred spaces where human knowledge was given permission to thrive.

A month later, I made the final stop on my journey of reconciliation. I awoke early and stepped into the brisk winter air. Traveling south on Interstate 45, I clicked on the CD console. Guitar arpeggios and pounding bongos rippled through the silence. The rhythms of bachata lifted my shoulders in time with the music and released the tension in my torso like the fingers of a masseuse. By the time I reached the junction of Interstate 10 East headed toward Mermentau, Louisiana, my spirit soared through the roof to Caribbean music with Angolan roots.

A weedy overgrowth had once demarcated the "colored" section of the cemetery; the well-groomed landscape now no longer announced its scars. The bucket of paint sloshed around my feet. As I crouched down at each concrete altar, my knees stung from kneeling on pebbles on wet grass. I slathered whitewash on the gravestones as my mind's eye shuffled through childhood memories to conjure up a living, breathing portrait of each of the deceased. The dappled light and shade of the afternoon sun danced through the gumwood trees, casting hieroglyphs on the gravestones. A yearning for something that I couldn't identify welled up deep within me. Is that what the old folks meant by ghosts? They line dance, twisting and twerking in pairs, leading all the way back to the dawn of time and wisdom.

NOTES

ACKNOWLEDGMENTS

INDEX

NOTES

Prologue

1 Eva K. F. Chan et al., "Human Origins in a Southern African Palaeo-Wetland and First Migrations," *Nature* 575, no. 7781 (October 2019): 185–189.

1. Our History

1 "African American Health: Creating Equal Opportunities for Health," Centers for Disease Control and Prevention, last updated July 3, 2017, https://www.cdc.gov/vitalsigns/aahealth/.

2 D. T. Lackland, "Racial Differences in Hypertension: Implications for High Blood Pressure Management," *American Journal of the Medical Sciences* 348, no. 2 (August 2014): 135–138.

3 Ibn Kaldūn, *The Muqaddimah: An Introduction to History,* trans. Franz Rosenthal, ed. N. J. Dawood, abridged ed. (Princeton: Princeton University Press, 1969). OR: Ibn Kaldūn, *The Muqaddimah: An Introduction to History,* trans. Franz Rosenthal, ed. N. J. Dawood, 3 vols. (Princeton: Princeton University Press, 1958). [or 3 vols., 2nd ed. 1967]

4 Leo Africanus, *Description de l'Afrique*, trans. and ed. Alexis Épaulard (Paris: Adrien Maissonneuve, 1956), 1:468–469.

5 See "Timbuktu Library Project," Hutchins Center for African & African American Research, Harvard University, https://hutchinscenter.fas.harvard.edu/timbuktu-library-project.

6 Naveena Kottoor, "How Timbuktu's Manuscripts Were Smuggled to Safety," *BBC News,* June 4, 2013, https://www.bbc.com/news/magazine-22704960.

7 Joseph Ki-Zerbo, *Alfred Diban: Premier chretien de Haute-Volta* (Paris: Cerf, 1983).

8 J. F. P. Hopkins and N. Levtzion, eds., *Corpus of Early Arabic Sources for West African History* (Cambridge: Cambridge University Press, 1981), 62–64.

9 Abd al-Rahman Sa'di, *Tarikh al-Sudan* (1655), Gallica, Manuscript B, Arabe 5256, Bibliothèque Nationale de France, Paris. My translation.

10 Sébastien Gasc, "Numismatics Data about the Islamic Conquest of the Iberian Peninsula," *Journal of Medieval Iberian Studies* 11, no. 3 (2019): 342–358.

11 Herodotus, *The Persian Wars,* vol. 1, trans. A. D. Godley, Loeb Classical Library 118 (Cambridge, MA: Harvard University Press, 1920), 4.196.

12 Lawrence Clayton, "Bartolomé de las Casas and the African Slave Trade," *History Compass* 7, no. 6 (2009): 1532.

13 Louis Jadin and Mirelle Dicorato, *Correspondance de Dom Afonso: roi du Congo, 1506–1543* (Brussels: Académie royale des sciences d'outre-mer, 1974). My translation.

14 Mungo Park, *Travels in the Interior Districts of Africa: Performed under the Direction and Patronage of the African Association, in the Years 1795, 1796, and 1797* (London: W. Bulmer and Co., 1799), 418.

15 John Matthews, *A Voyage to the River Sierra-Leone on the Coast of Africa; Containing an Account of the Trade and Productions of the Country, and of the Civil and Religious Customs and Manners of the People; in a Series of Letters to a friend in England* (London: B. White and Son, 1788).

16 Thomas W. Wilson, "History of Salt Supplies in West Africa and Blood Pressures Today," *The Lancet* 1, no. 8484 (5 April 1986): 784–786.

17 Thomas W. Wilson and Clarence E. Grim, "Biohistory of Slavery and Blood Pressure Differences in Blacks Today," *Hypertension* 17, 1 Suppl. I (1991): I 122–128; Graham A. MacGregor and Hugh E. de Wardener, *Salt, Diet and Health: Neptune's Poisoned Chalice: The Origins of High Blood Pressure* (Cambridge: Cambridge University Press, 1998), 5.

18 David M. Cutler, Roland G. Fryer, and Edward L. Glaeser, "Racial Differences in Life Expectancy: The Impact of Salt, Slavery, and Selection," unpublished ms., March 1, 2005.

19 W. J. Oliver, E. L. Cohen, and J. V. Neel, "Blood Pressure, Sodium Intake, and Sodium Related Hormones in the Yanomamo Indians, a 'No-Salt' Culture," *Circulation* 52 (July 1975): 146–151.

20 John H. Fountain, Jasleen Kaur, and Sarah L. Lappin, "Physiology, Renin Angiotensin System," StatPearls, NIH National Library of Medicine, last updated March 12, 2023, https://www.ncbi.nlm.nih.gov/books/NBK470410/; Pablo Nakagawa et al., "The Renin-Angiotensin System in the Central Nervous System and Its Role in Blood Pressure Regulation," *Current Hypertension Reports* 22, no. 7 (January 10, 2020): 1.

21 David J. Friedman et al., "Population-Based Risk Assessment of APOL1 on Renal Disease." *Journal of the American Society of Nephrology* 22, no. 11 (November 2011): 2098–2105; Pradeep Arora, "Chronic Kidney Disease (CKD)," Medscape, last updated May 26, 2023, https://emedicine.medscape.com/article/238798-overview#a1.

22 Mihail Zilbermint, Fady Hannah-Shmouni, and Constantine A. Stratakis, "Genetics of Hypertension in African Americans and Others of African Descent," *International Journal of Molecular Sciences* 20, no. 5 (2019): 1081.

23 James Anderson et al., "Racism: Eroding the Health of Black Communities," *JAAPA* 36, no. 5 (May 2023): 38–42; Courtney S. Thomas Tobin et al., "Discrimination, Racial Identity, and Hypertension among Black Americans across Young, Middle, and Older Adulthood," *Journals of Gerontology: Series B* 77, no. 11 (November 2022): 1990–2005; Osayande Agbonlahor et al., "Racial/Ethnic Discrimination and Cardiometabolic Diseases: A Systematic Review," *Journal of Racial and Ethnic Health Disparities* (2023): 1–25.

24 Rod S. Taylor et al., "Reduced Dietary Salt for the Prevention of Cardiovascular Disease: A Meta-Analysis of Randomized Controlled Trials (Cochrane Review)," *American Journal of Hypertension* 24, no. 8 (August 2011): 843–853; Linda J. Cobiac, Theo Vos, and J. Lennert Veerman, "Cost-Effectiveness of Interventions to Reduce Dietary Salt Intake," *Heart* 96 (2010): 1920–1925.

25 James DiNicolantonio, *The Salt Fix: Why the Experts Got It All Wrong—and How Eating More Might Save Your Life* (New York: Harmony Books, 2017).

26 See Kris Gunnars, "The Salt Myth—How Much Sodium Should You Eat Per Day?" Authority Nutrition: An Evidence-Based Approach, July 3, 2013, http://authoritynutrition.com/how-much-sodium-per-day (site discontinued), archived at: https://web.archive.org/web/20140528091911/http://authoritynutrition.com/how-much-sodium-per-day/.

27 Niels Graudal, Gesche Jürgens, Bo Baslund, and Michael H. Alderman, "Compared with Usual Sodium Intake, Low- and Excessive-Sodium Diets Are Associated with Increased Mortality: A Meta-analysis," *American Journal of Hypertension* 27, no. 9 (2014): 1129–1137.

28 Martin O'Donnell et al., "Urinary Sodium and Potassium Excretion, Mortality, and Cardiovascular Events," *New England Journal of Medicine* 371, no. 7 (2014): 612–623.

29 "Pour on the Salt? New Research Suggests More Is Okay," NBC News, August 13, 2014, http:/www.nbcnews.com/health/heart-health/pour-salt-new-research-suggests-more-ok-n179941.

30 Constance Hilliard, "Graudal et al. Article on Sodium Intake Should Include Ethnic Disclaimer," *American Journal of Hypertension* 27, no. 9 (September 2014): 1231.

31 Niels Graudal and Michael H. Alderman, "Response to 'Article on Sodium Intake Should Include Ethnic Disclaimer,'" *American Journal of Hypertension* 27, no. 9 (September 2014): 1232.

2. *The System*

1 Ann Gibbons, "How We Lost Our Diversity," *Science,* October 8, 2009, 2.

2 Ning Yu et al., "Larger Genetic Differences within Africans Than between Africans and Eurasians," *Genetics* 161, no. 1 (May 1, 2002): 269.

3 Nicholas G. Crawford et al., "Loci Associated with Skin Pigmentation Identified in African Populations," *Science* 358, no. 6365 (October 12, 2017); Meng Lin et al., "Rapid Evolution of a Skin-Lightening Allele in Southern African Khoesan," *Proceedings of the National Academy of Sciences* 115, no. 52 (December 26, 2018): 13324–13329.

4 Jason E. Lewis et al., "The Mismeasure of Science: Stephen Jay Gould versus Samuel George Morton on Skulls and Bias," *PLoS Biology* 9, no. 6 (June 2011).

5 John Hawks, "Accurate Depiction of Uncertainty in Ancient DNA Research: The Case of Neandertal Ancestry in Africa," *Journal of Social Archaeology* 21, no. 2 (June 2021): 179–196.

6 Lisa Weasel, "How Neanderthals Became White: The Introgression of Race into Contemporary Human Evolutionary Genetics," *American Naturalist* 200,

no. 1 (July 2022): 133. See also Jorge Rocha, "The Evolutionary History of Human Skin Pigmentation," *Journal of Molecular Evolution* 88 (2020): 77–87.

7 Sergey Nurk et al., "The Complete Sequence of a Human Genome," *Science* 376, no. 6588 (March 31, 2022): 53.

8 Joannella Morales et al., "A Standardized Framework for Representation of Ancestry Data in Genomics Studies, with Application to the NHGRI-EBI GWAS Catalog," *Genome Biology* 19, no. 21 (2018).

9 Ting Wang et al., "The Human Pangenome Project: A Global Resource to Map Genomic Diversity," *Nature* 604, no. 7906 (2022): 437–446.

10 Jeffrey C. Long, "Human Genetic Variation: The Mechanisms and Results of Microevolution," paper presented at the American Anthropological Association Annual Meeting, Chicago, November 21, 2003.

11 Jada Benn Torres and Rick A. Kittles, "The Relationship between 'Race' and Genetics in Biomedical Research," *Current Hypertension Reports* 9, no. 3 (2007): 196–201.

12 R. C. Lewontin, "The Apportionment of Human Diversity," *Evolutionary Biology* 6 (1972): 381–398.

13 Richard C. Lewontin, *The Genetic Basis of Evolutionary Change* (New York: Columbia University Press, 1974).

14 "Genetic Diversity: The Hidden Secret of Life," The Convention on Biological Diversity, May 31, 2021, https://www.cbd.int/article/genetic-diversity-the-hidden-secret-of-life#:~:text=All%20the%20biological%20data%20and,in%20climate%20and%20other%20stresses.

15 Minke Holwerda, "Pathogens in Permafrost: The Next Pandemic?" *Radboud Annals of Medical Students,* 21st ed. (December 2021): 26–29, https://www.ramsresearch.nl/wp-content/uploads/2021/12/21e-editie-RAMS-26-29.pdf.

16 G. D. Pule et al., "Beta-Globin Gene Haplotypes and Selected Malaria-Associated Variants among Black Southern African Populations," *Global Health, Epidemiology, and Genomics* 2 (2017).

17 Lon R. Cardon et al., "Genome-wide Association Study of 14,000 Cases of Seven Common Diseases and 3,000 Shared Controls," *Nature* 447, no. 7145 (2007): 661–678.

18 Brian Resnick, "Genetics Has Learned a Ton—Mostly about White People. That's a Problem," *Vox*, October 27, 2018, https://www.vox.com/science-and

-health/2018/10/22/17983568/dna-tests-precision-medicine-genetics-gwas-diversity-all-of-us.

19 F. S. Collins and V. A. McKusick, "Implications of the Human Genome Project for Medical Science," *Journal of the American Medical Association* 285, no. 5 (February 2001): 540–544; "dbSNP's Human Build 150 Has Doubled the Amount of RefSNP Records!" National Institute of Health, National Library of Medicine, National Center for Biotechnology Information, NCBI Insights, May 8, 2017, https://ncbiinsights.ncbi.nlm.nih.gov/2017/05/08/dbsnps-human-build-150-has-doubled-the-amount-of-refsnp-records/.

20 Oliver Mayo, "The Rise and Fall of the Common Disease–Common Variant (CD–CV) Hypothesis: How the Sickle Cell Disease Paradigm Led Us All Astray (Or Did It?)," *Twin Research and Human Genetics* 10, no. 6 (December 2007): 793–804.

21 Brendan Maher, "Personal Genomes: The Case of the Missing Heritability," *Nature* 456 (November 6, 2008): 18–21.

22 Sarah A. Tishkoff et al., "Convergent Adaptation of Human Lactose Persistence in Africa and Europe," *Nature Genetics* 39, no. 1 (January 2007): 31–40.

23 Rachel M. Sherman et al., "Assembly of a Pan-Genome from Deep Sequencing of 910 Humans of African Descent," *Nature Genetics* 51, no. 1 (2019): 30–35.

24 Sharon Begley, "Buffalo Gave Us Spicy Wings and the 'Book of Life.' Here's Why That's Undermining Personalized Medicine," *Stat,* March 11, 2019, https://www.statnews.com/2019/03/11/human-reference-genome-shortcomings/.

25 Hassan Ashktorab et al., "Racial Disparity in Gastrointestinal Cancer Risk," *Gastroenterology* 153, no. 4 (October 2017): 910–923.

26 Stanley E. Hooker, Jr., et al., "Genetic Ancestry Analysis Reveals Misclassification of Commonly Used Cancer Cell Lines," *Cancer Epidemiology, Biomarkers & Prevention* 28, no. 6 (June 2019): 1003–1009.

27 Harriet A. Washington, *Medical Apartheid: The Dark History of Medical Experimentation on Black Americans from Colonial Times to the Present* (New York: Anchor, 2008), 62–63.

28 Marie Jenkins Schwartz, *Birthing a Slave: Motherhood and Medicine in the Antebellum South* (Cambridge, MA: Harvard University Press, 2006).

29 Howard Markel, "Appendix 6: Scientific Advances and Social Risks; Historical Perspectives of Genetic Screening Programs for Sickle Cell Disease, Tay-Sachs Disease, Neural Tube Defects, and Down Syndrome, 1970–1997," in *Promoting Safe and Effective Genetic Testing in the United States,* ed. Neil A. Holzman and Michael Watson, Final Report of the Task Force on Genetic Testing, National Human Genome Research Institute, National Institutes of Health, September 1997, https://biotech.law.lsu.edu/research/fed/tfgt/appendix6.htm.

30 Michael LaForgia and Jennifer Valentino-DeVries, "How a Genetic Trait in Black People Can Give the Police Cover," *New York Times,* May 15, 2021.

31 David Richardson, "Shipboard Revolts, African Authority, and the Atlantic Slave Trade," *William and Mary Quarterly* 58, no. 1 (2001): 69–92.

32 Felicia Gomez, Jibril Hirbo, and Sarah A. Tishkoff, "Genetic Variation and Adaptation in Africa: Implications for Human Evolution and Disease," *Cold Spring Harbor Perspectives in Biology* 6, no. 7 (July 2014).

33 Amit V. Khera et al., "Genome-Wide Polygenic Scores for Common Diseases Identify Individuals with Risk Equivalent to Monogenic Mutations," *Nature Genetics* 50, no. 9 (September 2018): 1219–1224.

34 Alicia R. Martin et al., "Current Clinical Use of Polygenic Scores Will Risk Exacerbating Health Disparities," *Nature Genetics* 51 (April 2019): 584–591.

35 Kathryn Paige Harden, *The Genetic Lottery: Why DNA Matters for Social Equality* (Princeton, NJ: Princeton University Press, 2021).

36 Kevin A. Bird, "No Support for the Hereditarian Hypothesis of the Black–White Achievement Gap Using Polygenic Scores and Tests for Divergent Selection," *American Journal of Physical Anthropology* 175 (2021): 465.

37 L. Duncan et al., "Analysis of Polygenic Risk Score Usage and Performance in Diverse Human Populations," *Nature Communications* 10, no. 3328 (2019).

38 "Educational Attainment," United States Census Bureau, https://www.census.gov/topics/education/educational-attainment.html#:~:text=Educational%20attainment%20refers%20to%20the,Publications, accessed April 2022.

39 Aysu Okbay et al., "Polygenic Prediction of Educational Attainment within and between Families from Genome-Wide Association Analyses in 3 Million Individuals," *Nature Genetics* 54 (April 2022): 437–449.

40 Joseph L. Graves, Jr., *The Emperor's New Clothes: Biological Theories of Race at the Millennium* (New Brunswick, NJ: Rutgers University Press, 2001); Joel Z. Garrod, "A Brave Old World: An Analysis of Scientific Racism and BiDil®," *McGill Journal of Medicine* 9, no. 1 (January 2006): 54–60.

41 Josiah C. Nott (Josiah Clark) and George R. Gliddon (George Robins), *Types of Mankind: Or, Ethnological Researches Based upon the Ancient Monuments, Paintings, Sculptures, and Crania of Races, and upon Their Natural, Geographical, Philological and Biblical History* (Philadelphia: J. B. Lippincott, Grambo & Co., 1854).

42 Francis Galton, *Inquiries into Human Faculty and Its Development* (London: MacMillan, 1883), 25.

43 Charles Darwin, *The Descent of Man, and Selection in Relation to Sex* (London: John Murray, 1871).

44 A. Smith Woodward, "I.—Note on the Piltdown Man (Eoanthropus Dawsoni)," *Geological Magazine* 10, no. 10 (1913): 433–434.

45 William L. Straus, Jr., "The Great Piltdown Hoax," *Science* 119, no. 3087 (1954): 265–269.

46 Alex Ross, "How American Racism Influenced Hitler," *New Yorker,* April 30, 2018.

47 David Starr Jordan, *Blood of a Nation: A Study of the Decay of Races through Survival of the Unfit* (Boston: American Unitarian Association, 1902).

48 Edwin Black, *War against the Weak: Eugenics and America's Campaign to Create a Master Race,* expanded ed. (Washington, DC: Dialog Press, 2012).

49 Paul A. Lombardo, "'The American Breed': Nazi Eugenics and the Origins of the Pioneer Fund," *Albany Law Review* 65, no. 3 (2001): 743.

50 William Tucker, *The Funding of Scientific Racism: Wickliffe Draper and the Pioneer Fund* (Champaign: University of Illinois Press, 2007).

51 Richard J. Herrnstein and Charles Murray, *The Bell Curve: Intelligence and Class Structure in American Life* (New York: Free Press, 1994).

52 Rushton was quoted in a 1994 interview with *Rolling Stone* as saying, "It's a trade-off, more brains or more penis. You can't have everything." Rushton claims the statement is a fabrication. Adam Miller, "Professors of Hate," *Rolling Stone,* October 20, 1994, 106–114.

53 Paul Selvin, "The Raging Bull of Berkeley," *Science* 251, no. 4992 (January 25, 1991): 368.

54 Constance Hilliard, *Straightening the Bell Curve: How Stereotypes about Black Masculinity Drive Research on Race and Intelligence* (Washington, DC: Potomac Books, 2012), 69–76.

55 David Epstein, *The Sports Gene: Inside the Science of Extraordinary Athletic Performance* (New York: Current, 2013), 144.

56 Kip D. Zimmerman et al., "Genetic Landscape of Gullah African Americans," *American Journal of Physical Anthropology* 175, no 4 (2021): 905–919.

57 Tanda Murray et al., "African and Non-African Admixture Components in African Americans and an African Caribbean Population," *Genetic Epidemiology* 34, no. 6 (2010): 561–568.

58 Chen Zhu et al., "Hybrid Marriages and Phenotypic Heterosis in Offspring: Evidence from China," *Economics & Human Biology* 29 (May 2018): 102–114.

59 Andrea Manica et al., "The Effect of Ancient Population Bottlenecks on Human Phenotypic Variation," *Nature* 448, no. 7151 (July 2007): 346–348.

60 Richard Lynn and Tatu Vanhanen, "National IQ and Economic Development: A Study of Eighty-One Nations," *Mankind Quarterly* 41, no. 4 (June 2001): 415–435.

61 Angela Saini, *Superior: The Return of Race Science* (Boston: Beacon Press, 2019).

62 Alexander P. Burgoyne and David Z. Hambrick, "Intelligence and the DNA Revolution," *Scientific American Mind* 28, no. 6 (November/December 2017): 12–14.

63 Suzanne Sniekers et al., "Genome-Wide Association Meta-Analysis of 78,308 Individuals Identifies New Loci and Genes Influencing Human Intelligence," *Nature Genetics* 49, no. 7 (July 2017): 1107–1112.

64 Alexander P. Burgoyne et al., "Differential and Experimental Approaches to Studying Intelligence in Humans and Non-Human Animals," *Learning and Motivation* 72 (November 2020).

65 Monika Pronczuk and Koba Ryckewaert, "A Racist Researcher, Exposed by a Mass Shooting," *New York Times,* June 9, 2022.

66 Lucia A. Hindorff, Elizabeth M. Gillanders, and Teri A. Manolio, "Genetic Architecture of Cancer and Other Complex Diseases: Lessons Learned and Future Directions," *Carcinogenesis* 32, no. 7 (July 2011): 945–954.

67 "Global Personalized Medicine (PM) Market Anticipated to Near $3.2 Trillion by 2025–PM Therapeutics Is Projected to Register the Fastest CAGR,"

Globe Newswire, August 21, 2019, https://www.globenewswire.com/news-release/2019/08/21/1904824/0/en/Global-Personalized-Medicine-PM-Market-Anticipated-to-Near-3-2-Trillion-by-2025-PM-Therapeutics-is-Projected-to-Register-the-Fastest-CAGR.html.

3. *Our Health*

1 There are, however, certain dwarf breeds of cattle, bred by Fulbe nomadic pastoralists, that carry some level of natural immunity to the disease, but the percentage of such breeds is small.

P. H. Clausen, I. Sidibe, I. Kabore, and B. Bauer, "Development of Multiple Drug Resistance of Trypanosoma Congolense in Zebu Cattle under High Natural Tsetse Fly Challenge in the Pastoral Zone of Samorogouan, Burkina Faso," *Acta Tropica* 51, no. 3–4 (1992): 229–236.

2 Roger Martin Djoumessi Zebaze and Ego Seeman, "Epidemiology of Hip and Wrist Fractures in Cameroon, Africa," *Osteoporosis International* 14, no. 4 (June 2003): 301–305.

3 Robert P. Heaney, "Ethnicity, Bone Status, and the Calcium Requirement," *Nutrition Research* 22, nos. 1–2 (January/February 2002): 153–178.

4 Melissa S. Putnam et al., "Differences in Skeletal Microarchitecture and Strength in African-American and White Women," *Journal of Bone and Mineral Research* 28, no. 10 (2013): 2177–2185.

5 Marlene Chakhtoura and Ghada El-Hajj Fuleihan, "Treatment of Hypercalcemia of Malignancy," *Endocrinology and Metabolism Clinics* 50, no. 4 (2021): 781–792.

6 Tao Na et al., "The A563T Variation of the Renal Epithelial Calcium Channel TRPV5 among African Americans Enhances Calcium Influx," *American Journal of Physiology–Renal Physiology* 296, no. 5 (May 2009): F1042–F1051; Yoshiro Suzuki et al., "Gain-of-Function Haplotype in the Epithelial Calcium Channel TRPV6 Is a Risk Factor for Renal Calcium Stone Formation," *Human Molecular Genetics* 17, no. 11 (June 2008): 1613–1618.

7 Na et al., "The A563T Variation."

8 David A. Hughes et al., "Parallel Selection on TRPV6 in Human Populations," *PLOS One* 3, no. 2 (February 2008).

9 Constance B. Hilliard, "High Osteoporosis Risk among East Africans Linked to Lactase Persistence Genotype," *BoneKEy Reports* 5 (2016): 803.

10 V. Lehen'kyi et al., "TRPV6 Channel Controls Prostate Cancer Cell Proliferation via Ca^{2+}/NFAT-Dependent Pathways," *Oncogene* 26 (2007): 7380–7385.

11 Constance Hilliard, "Ecological Model Links Proto-Oncogene to High Incidence of Metastatic Cancers in African-Americans," *Journal of Cancer Research & Therapy* 6, no. 5 (2018): 37–40.

12 Patricia A. Frances-Lyon et al., "TRPV6 as a Putative Genomic Susceptibility Locus Influencing Racial Disparities in Cancer," *Cancer Prevention Research* 13, no. 5 (May 2020): 423–427.

13 Na et al., "The A563T Variation."

14 Ismail Kaddour-Djebbar et al., "Specific Mitochondrial Calcium Overload Induces Mitochondrial Fission in Prostate Cancer Cells," *International Journal of Oncology* 36, no. 6 (June 2010): 1437–1444.

15 Siddhartha Yaday, "Impact of *BRCA* Mutation Status on Survival of Women with Triple-Negative Breast Cancer," *Clinical Breast Cancer* 18, no. 5 (October 2018): e1229–e1235.

16 Jian-Hao Liu et al., "Trichostatin A Induces Autophagy in Cervical Cancer Cells by Regulating the PRMT5-STC1-TRPV6-JNK Pathway," *Pharmacology* 106, nos. 1–2 (February 2021): 60–69; Lingying Wu et al., "A Phase Ia/Ib Study of CBP-1008, a Bispecific Ligand Drug Conjugate, in Patients with Advanced Solid Tumors," *Journal of Clinical Oncology* 40, no. 16 suppl. (2022): 3000.

17 Katarzyna Pogoda et al., "Effects of *BRCA* Germline Mutations on Triple-Negative Breast Cancer Prognosis," in "BRCA Mutations in Cancer: Implications for Tumor Biology, Surveillance, and Treatment," ed. Angela Toss, special issue, *Journal of Oncology* 2020 (November 2020); Elizabeth Pratt, "Triple-Negative Breast Cancer and BRCA1 Mutation: What Is the Link?" Medical News Today, February 22, 2022, https://www.medicalnewstoday.com/articles/triple-negative-breast-cancer-and-brca1-mutation?utm_source=ReadNext.

18 Qin Wu et al., "GLUT1 Inhibition Blocks Growth of RB1-Positive Triple Negative Breast Cancer," *Nature Communications* 11, no. 4205 (2020).

19 Thorsten Kessler et al., "TRPV6 Alleles Do Not Influence Prostate Cancer Progression," *BMC Cancer* 9, no. 380 (2009).

20 Loren Saulsberry et al., "Precision Oncology: Directing Genomics and Pharmacogenomics toward Reducing Cancer Inequities," *Cancer Cell* 39, no. 6 (June 2021): 730–733.

21 Kelvin Li et al., "The Good, the Bad, and the Ugly of Calcium Supplementation: A Review of Calcium Intake on Human Health," *Clinical Interventions in Aging* 13 (2018): 2446. See also Landing M. A. Jarjou et al., "Effect of Calcium Supplementation in Pregnancy on Maternal Bone Outcomes in Women with a Low Calcium Intake," *American Journal of Clinical Nutrition* 92, no. 2 (August 2010): 450–457.

22 Tonya Russell, "Mortality Rate for Black Babies Is Cut Dramatically When Black Doctors Care for Them after Birth, Researchers Say," *Washington Post*, January 13, 2021.

23 "Black Women Over Three Times More Likely to Die in Pregnancy, Postpartum Than White Women, New Research Finds," Population Reference Bureau, December 6, 2021, https://www.prb.org/resources/black-women-over-three-times-more-likely-to-die-in-pregnancy-postpartum-than-white-women-new-research-finds/.

24 Alessia Mammaro et al., "Hypertensive Disorders of Pregnancy," *Journal of Prenatal Medicine* 3, no. 1 (January–March 2009): 1–5.

25 Donna L. Hoyert, "Maternal Mortality Rates in the United States, 2021," Health E-Stats, Centers for Disease Control and Prevention, March 2023, https://www.cdc.gov/nchs/data/hestat/maternal-mortality/2021/maternal-mortality-rates-2021.htm.

26 New York Times Editorial Board, "Easing the Dangers of Childbirth for Black Women," *New York Times*, April 20, 2018.

27 James R. Sowers et al., "Postpartum Abnormalities of Carbohydrate and Cellular Calcium Metabolism in Pregnancy Induced Hypertension," *American Journal of Hypertension* 6, no. 4 (April 1993): 302–307.

28 Lawrence M. Resnick, "The Role of Dietary Calcium in Hypertension: A Hierarchal Overview," *American Journal of Hypertension* 12, no. 1 (January 1999): 99–112.

29 Safiya I. Richardson et al., "Salt Sensitivity: A Review with a Focus on Non-Hispanic Blacks and Hispanics," *Journal of the American Society of Hypertension* 7, no. 2 (March–April 2013): 170–179.

30 E.V. Souza et al., "Aspirin Plus Calcium Supplementation to Prevent Superimposed Preeclampsia: A Randomized Trial," *Brazilian Journal of Medical and Biological Research* 47, no. 5 (May 2014): 419–425.

31 John F. Aloia, "African Americans, 25-Hydroxyvitamin D, and Osteoporosis: A Paradox," *American Journal of Clinical Nutrition* 88, no. 2 (August 2008): 545S–550S.

32 Heidi J. Kalkwarf et al., "The Bone Mineral Density in Childhood Study: Bone Mineral Content and Density According to Age, Sex, and Race," *Journal of Clinical Endocrinology & Metabolism* 92, no. 6 (June 2007): 2087–2099; William D. Leslie, "Ethnic Differences in Bone Mass—Clinical Implications," *Journal of Clinical Endocrinology & Metabolism* 97, no. 12 (December 2012): 4329–4340; Chao Tian et al., "A Genomewide Single-Nucleotide–Polymorphism Panel with High Ancestry Information for African American Admixture Mapping," *American Journal of Human Genetics* 79, no. 4 (October 2006): 640–649.

33 Constance B. Hilliard, "High Osteoporosis Risk among East Africans Linked to Lactase Persistence Genotype," *BoneKEy Reports* 5, no. 803 (June 2016).

34 Judith A. Carney, *Black Rice: The African Origins of Rice Cultivation in the Americas* (Cambridge, MA: Harvard University Press, 2002).

35 Greg Timmons, "How Slavery Became the Economic Engine of the South," History Channel, March 6, 2018, https://www.history.com/news/slavery-profitable-southern-economy.

36 Sarah A. Tishkoff et al., "The Genetic Structure and History of Africans and African Americans," *Science* 324, no. 5930 (April 2009): 1035–1044.

37 Jennifer Jensen Wallach, *How America Eats: A Social History of US Food and Culture* (Lanham, MD: Rowman & Littlefield, 2013).

38 Jennifer J. Wallach, *Every Nation Has Its Dish: Black Bodies & Black Foods in Twentieth-Century America* (Chapel Hill: University of North Carolina Press, 2019).

39 Julia Moskin, "Is It Southern Food, or Soul Food," *New York Times*, August 7, 2018.

40 Barry M. Popkin et al., "A Comparison of Dietary Trends among Racial and Socioeconomic Groups in the United States," *New England Journal of Medicine* 335, no. 10 (1996): 716–720.

41 Marie T. Ruel, "Operationalizing Dietary Diversity: A Review of Measurement Issues and Research Priorities," *Journal of Nutrition* 133, no. 11 (November 2003): 3911S–3926S.

42 Maria C. de Oliveira Otto et al., "Dietary Diversity: Implications for Obesity Prevention in Adult Populations; A Science Advisory from the American Heart Association," *Circulation* 138, no. 11 (September 2018): e160–e168.

43 BIDMC Communications, "DASH Diet's Impact Differs Based on Race and Gender," *Harvard Gazette,* November 30, 2022, https://news.harvard.edu/gazette/story/2022/11/dash-diet-offers-even-more-benefits-for-black-adults-and-women/; for the full study, see Sun Young Jeong et al., "Effects of Diet on 10-Year Atherosclerotic Cardiovascular Disease Risk (from the DASH Trial)," *American Journal of Cardiology* 187 (January 2023): 10–17.

44 Aries Chavira et al., "The Microbiome and Its Potential for Pharmacology," in *Concepts and Principles of Pharmacology: 100 Years of the Handbook of Experimental Pharmacology,* ed. James E. Barrett, Clive P. Page, and Martin C. Michael (Cham, Switzerland: Springer, 2019), 301.

45 Vinod K. Gupta, Sandip Paul, and Chitra Dutta, "Geography, Ethnicity or Subsistence-Specific Variations in Human Microbiome Composition and Diversity," *Frontiers in Microbiology* 8, no. 1162 (June 2017).

4. *The Algorithm*

1 Jessica Guynn, "Facebook Apologizes after Mislabeling Video of Black Men as 'Primates,'" *USA Today,* September 3, 2021; Kashmir Hill, "Wrongfully Accused by an Algorithm," *New York Times,* June 24, 2020.

2 Carolyn Y. Johnson, "Racial Bias in a Medical Algorithm Favors White Patients Over Sicker Black Patients," *Washington Post,* October 24, 2019.

3 Lauren A. Eberly et al., "Identification of Racial Inequities in Access to Specialized Inpatient Heart Failure Care at an Academic Medical Center," *Circulation: Heart Failure* 12, no. 11 (2019): e006214.

4 Darshali A. Vyas, Leo G. Eisenstein, and David S. Jones, "Hidden in Plain Sight—Reconsidering the Use of Race Correction in Clinical Algorithms," *New England Journal of Medicine* 383, no. 9 (August 2020): 874–882.

5 Jennifer Tsai, "Jordan Crowley Would Be in Line for a Kidney—If He Were Deemed White Enough," *Slate,* June 27, 2021.

6 Katarzyna Bryc et al., "The Genetic Ancestry of African Americans, Latinos, and European Americans across the United States," *American Journal of Human Genetics* 96, no. 1 (January 2015): 37–53.

7 Michael Yudell et al., "Taking Race out of Human Genetics," *Science* 351, no. 6273 (February 2016): 564–565.

8 Joy Hsu, Kirsten L. Johansen, Chi-yuan Hsu, George A. Kaysen, and Glenn M. Chertow, "Higher Serum Creatinine Concentrations in Black Patients with Chronic Kidney Disease: Beyond Nutritional Status and Body Composition," *Clinical Journal of the American Society of Nephrology* 3, no. 4 (2008): 992–997.

9 Nwamaka Denise Eneanya, Wei Yang, and Peter Philip Reese, "Reconsidering the Consequences of Using Race to Estimate Kidney Function," *Journal of the American Medical Association* 322, no. 2 (July 2019): 113–114.

10 Ana Prohaska et al., "Human Disease Variation in the Light of Population Genomics," *Cell* 177, no. 1 (March 2019): 115–131.

11 National Academies of Sciences, Engineering, and Medicine, *Using Population Descriptors in Genetics and Genomics Research: A New Framework for an Evolving Field* (Washington, DC: The National Academies Press, 2023).

12 Bessie A. Young et al., "Clinical Genetic Testing for APOL1: Are We There Yet?" *Seminars in Nephrology* 37, no. 6 (November 2017): 552–557.

5. Ancestral Genomics

1 Patricia Balaresque et al., "A Predominantly Neolithic Origin for European Paternal Lineages," *PLOS Biology* 8, no. 1 (January 2010): e1000285; Natalie M. Myres et al., "A Major Y-Chromosome Haplogroup R1b Holocene Era Founder Effect in Central and Western Europe," *European Journal of Human Genetics* 19, no. 1 (2011): 95–101.

2 Joseph L. Graves, Jr., et al., "Evolutionary Science as a Method to Facilitate Higher Level Thinking and Reasoning in Medical Training," *Evolution, Medicine, and Public Health* 2016, no. 1 (January 2016): 358–368.

3 Otto Juettner, *Modern Physio-Therapy: A System of Drugless Therapeutic Methods* (Cincinnati, OH: Harvey, 1906), 294.

4 Itai Bavli and David S. Jones, "Race Correction and the X-ray Machine—The Controversy over Increased Radiation Doses for Black Americans in 1968," *New England Journal of Medicine* 387, no. 10 (September 2022): 947–952.

5 National Center for Radiological Health, "Patient Characteristics in the Determination of Radiographic Exposure Factors," *Journal of the National Medical Association* 61 (1969): 282–283; Bayli and Jones, "Race Correction."

6 Akinyemi Oni-Orisan et al., "Embracing Genetic Diversity to Improve Black Health," *New England Journal of Medicine* 384, no. 12 (March 2021): 1163–1167.

7 Michael C. Campbell and Sarah A. Tishkoff, "African Genetic Diversity: Implications for Human Demographic History, Modern Human Origins, and Complex Disease Mapping," *Annual Review of Genomics and Human Genetics* 9 (September 2008): 403–433.

8 Brian W. Kunkle et al., "Novel Alzheimer Disease Risk Loci and Pathways in African American Individuals Using the African Genome Resources Panel: A Meta-Analysis," *JAMA Neurology* 78, no. 1 (2021): 102–113.

9 Stephanie J. Monroe, "Discussing Inconvenient Truths about the Lack of Generalizability of Alzheimer's Research to Minoritized Populations," *Alzheimer's & Dementia: Translational Research & Clinical Interventions* 9, no. 3 (2023), e12406; J. Cummings et al., "Lecanemab: Appropriate Use Recommendations," *Journal of Prevention of Alzheimer's Disease* 10, no. 3 (2023): 362–377.

10 Eric Trépo and Luca Valenti, "Update on NAFLD Genetics: From New Variants to the Clinic," *Journal of Hepatology* 72, no. 6 (October 2020): 1196–1209; Valerie D. Myers et al., "Association of Variants in BAG3 with Cardiomyopathy Outcomes in African American Individuals," *JAMA Cardiology* 3, no. 10 (2018): 929–938; Mihail Zilbermint, Fady Hannah-Shmouni, and Constantine A. Stratakis, "Genetics of Hypertension in African Americans and Others of African Descent," *International Journal of Molecular Sciences* 20, no. 5 (2019): 1081; Héctor Díaz-Zabala et al., "Evaluating Breast Cancer Predisposition Genes in Women of African Ancestry," *Genetics in Medicine* 24, no. 7 (July 2022): 1468–1475; Guanyi Zhang et al., "A Recurrent ADPRHL1 Germline Mutation Activates PARP1 and Confers Prostate Cancer Risk in African American Families," *Molecular Cancer Research* 20, no. 12 (December 2022): 1776–1784; William L. Lowe, Jr., et al., "Genetics of Gestational Diabetes Mellitus and Maternal Metabolism," *Current Diabetes Reports* 16, no. 2 (January 2016): 1–10; Sara Clohisey and Pratik Sinha, "Genome-Wide Association Studies in ARDS: SNPing the Tangled Web of Heterogeneity," *Intensive Care Medicine* 47, no. 7 (2021): 782–785; Salim S. Virani et al., "Heart Disease and Stroke Statistics—2021 Update: A Report from the American Heart Association," *Circulation* 143, no. 8 (2021): e254–e743.

11 Genome Wide Association Studies (GWAS) Diversity Monitor, "Questions and Answers," https://gwasdiversitymonitor.com/qandas; Jacob A. Tennessen

et al., "Evolution and Functional Impact of Rare Coding Variation from Deep Sequencing of Human Exomes," *Science* 337, no. 6090 (2012): 64–69.

12 Lauren A. Wise, Julie R. Palmer, David Reich, Yvette C. Cozier, and Lynn Rosenberg, "Hair Relaxer Use and Risk of Uterine Leiomyomata in African-American Women," *American Journal of Epidemiology* 175, no. 5 (2012): 432–440.

13 "Hair Straightening Chemicals Associated with Higher Uterine Cancer Risk," National Institutes of Health, news release, October 17, 2022, https://www.nih.gov/news-events/news-releases/hair-straightening-chemicals-associated-higher-uterine-cancer-risk.

14 Catlin Nalley, "Initial Data Shows Potential of Gene Therapy for Sickle Cell Anemia," *Oncology Times* 41, no. 3 (February 2019): 15–16; Haydar Frangoul et al., "CRISPR-Cas9 Gene Editing for Sickle Cell Disease and β-Thalassemia," *New England Journal of Medicine* 384, no. 3 (January 2021): 252–260; Julie Kanter and Corey Falcon, "Gene Therapy for Sickle Cell Disease: Where We Are Now?" *Hematology* 2021, no. 1 (December 2021): 174–180.

15 Arjun K. Manrai et al., "Genetic Misdiagnoses and the Potential for Health Disparities," *New England Journal of Medicine* 375, no. 7 (August 2016): 655–665.

16 Jonathan Kahn, *Race in a Bottle: The Story of BiDil and Racialized Medicine in a Post-Genomic Age* (New York: Columbia University Press, 2013).

17 Sheldon Krimsky, "The Short Life of a Race Drug," *Lancet* 379, no. 9811 (January 2012): 114–115.

18 Linda Adeles, "Race-Based Prescribing for Black People with High Blood Pressure Shows No Benefit," UCSF, Patient Care, January 18, 2022. https://www.ucsf.edu/news/2022/01/422151/race-based-prescribing-black-people-high-blood-pressure-shows-no-benefit.

19 Alyson L. Dickson et al., "Race, Genotype, and Azathioprine Discontinuation: A Cohort Study," *Annals of Internal Medicine* 175, no. 8 (2022): 1092–1099.

20 Merlin C. Thomas et al., "The Association between Dietary Sodium Intake, ESRD, and All-Cause Mortality in Patients with Type 1 Diabetes," *Diabetes Care* 34, no. 4 (2011): 861–866; Martin O'Donnell, Andrew Mente, and Salim Yusuf, "Sodium Intake and Cardiovascular Health," *Circulation Research* 116, no. 6 (March 2015): 1046–1057; Niels Albert Graudal, Thorbjørn Hubeck-Graudal, and Gesche Jurgens, "Effects of Low Sodium Diet versus High Sodium Diet

on Blood Pressure, Renin, Aldosterone, Catecholamines, Cholesterol, and Triglyceride," *Cochrane Database of Systematic Reviews,* no. 12 (2020).

21 Janet Woodcock and Susan T. Mayne, "To Improve Nutrition and Reduce the Burden of Disease, FDA Issues Food Industry Guidance for Voluntarily Reducing Sodium in Processed and Packaged Foods," US Food and Drug Administration, October 13, 2021, https://www.fda.gov/news-events/press-announcements/improve-nutrition-and-reduce-burden-disease-fda-issues-food-industry-guidance-voluntarily-reducing.

22 Safiya I. Richardson et al., "Salt Sensitivity: A Review with a Focus on Non-Hispanic Blacks and Hispanics," *Journal of the American Society of Hypertension* 7, no. 2 (March–April 2013): 170–179.

23 Pierrick Uzureau et al., "Mechanism of *Trypanosoma brucei gambiense* Resistance to Human Serum," *Nature* 501, no. 7467 (September 2013): 430–434.

24 Camille E. Powe et al., "Vitamin D–Binding Protein and Vitamin D Status of Black Americans and White Americans," *New England Journal of Medicine* 369, no. 21 (November 2013): 1991–2000.

25 Constance B. Hilliard, "An Overlooked African Gene Variant Linked to the Calcium Selective Channel TRPV6: A Mini-Review," *Gene* 872, no. 1 (July 2023): 147429.

26 David A. Hughes et al., "Parallel Selection on TRPV6 in Human Populations," *PLoS One* 3, no. 2 (February 2008): e1686; Mark J. Bolland et al., "Calcium Intake and Risk of Fracture: Systematic Review," *BMJ* 351 (2015): h4580; Robert P. Heaney, "Low Calcium Intake among African Americans: Effects on Bones and Body Weight," *Journal of Nutrition* 136, no. 4 (April 2006): 1095–1098; "Age-Adjusted U.S. Death Rates and Trends for the Top 15 Cancer Sites: Both Sexes by Race/Ethnicity," National Cancer Institute, Surveillance, Epidemiology, and End Results Program, https://seer.cancer.gov/archive/csr/1975_2011/browse_csr.php?sectionSEL=1&pageSEL=sect_01_table.27.

27 Jose D. Debes et al., "Inverse Association between Prostate Cancer and the Use of Calcium Channel Blockers," *Cancer Epidemiology, Biomarkers & Prevention* 13, no. 2 (February 2004): 255–259.

28 Kevin R. Loughlin, "Calcium Channel Blockers and Prostate Cancer," *Urologic Oncology: Seminars and Original Investigations* 32, no. 5 (July 2014): 537–538.

29 Chris V. Bowen et al., "*In Vivo* Detection of Human TRPV6-Rich Tumors with Anti-Cancer Peptides Derived from Soricidin," *PLoS One* 8, no. 3 (March 2013): e58866.

30 John M. Stewart, "TRPV6 as a Target for Cancer Therapy," *Journal of Cancer* 11, no. 2 (2020): 374–387.

31 Yuan Jiang et al., "Lidocaine Inhibits the Invasion and Migration of TRPV6-Expressing Cancer Cells by TRPV6 Downregulation," *Oncology Letters* 12, no. 2 (August 2016): 1164–1170.

32 Stewart, "TRPV6 as a Target for Cancer Therapy."

33 David E. Winickoff and Osagie K. Obasogie, "Race-Specific Drugs: Regulatory Trends and Public Policy," *Trends in Pharmacological Sciences* 29, no. 6 (June 2008): 277–279.

34 Brian Kennedy, Alec Tyson, and Cary Funk, "Americans Value U.S. Role as Scientific Leader, but 38% Say Country Is Losing Ground Globally," Pew Research Center, October 25, 2022, https://www.pewresearch.org/science/2022/10/25/americans-value-u-s-role-as-scientific-leader-but-38-say-country-is-losing-ground-globally/.

35 Meggan J. Lee et al., "'If You Aren't White, Asian or Indian, You Aren't an Engineer': Racial Microaggressions in STEM Education," *International Journal of STEM Education* 7, no. 48 (2020).

36 David M. Quinn and North Cooc, "Science Achievement Gaps by Gender and Race/Ethnicity in Elementary and Middle School: Trends and Predictors," *Educational Researcher* 44, no. 6 (August–September 2015): 336–346.

37 Yingyi Ma and Yan Liu, "Race and STEM Degree Attainment," *Sociology Compass* 9, no. 7 (July 2015): 609–618.

38 "Statistics: Inmate Race," Federal Bureau of Prisons, last updated June 10, 2023, https://www.bop.gov/about/statistics/statistics_inmate_race.jsp.

39 Samuel R. Gross, Maurice Possley, and Klara Stephens, "Race and Wrongful Convictions in the United States," National Registry of Exonerations, March 7, 2017, https://repository.law.umich.edu/other/122/.

40 Agnes Constante, "Two-Thirds of Asian American Health, Food Workers Fighting COVID-19 Are Immigrants, Report Says," *NBC News*, June 8, 2020,

https://www.nbcnews.com/news/asian-america/two-thirds-asian-american-health-food-workers-fighting-covid-19-n1224721.

41 Nina Hollfelder et al., "The Deep Population History in Africa," *Human Molecular Genetics* 30, no. R1 (March 2021): R2–R10.

42 "International Consortium Announces the 1000 Genomes Project," National Institutes of Health, News Releases, January 22, 2008, https://www.nih.gov/news-events/news-releases/international-consortium-announces-1000-genomes-project.

43 Xiangqun Zheng-Bradley et al., "Alignment of 1000 Genomes Project Reads to Reference Assembly GRCh38," *GigaScience* 6, no. 7 (July 2017): 1–8.

44 The International HapMap 3 Consortium, "Integrating Common and Rare Genetic Variation in Diverse Human Populations," *Nature* 467, no. 7311 (September 2010): 52–58.

45 Omolola A. Adedokun et al., "Research Skills and STEM Undergraduate Research Students' Aspirations for Research Careers: Mediating Effects of Research Self-Efficacy," *Journal of Research in Science Teaching* 50, no. 8 (October 2013): 940–951.

46 Linda Nordling, "Africa Analysis: Lawsuit Offers Lessons for Alliances," SciDev.Net, August 14, 2014, https://www.scidev.net/sub-saharan-africa/column/africa-analysis-alliances/.

47 Hannes Toivanen and Branco Ponomariov, "African Regional Innovation Systems: Bibliometric Analysis of Research Collaboration Patterns 2005–2009," *Scientometrics* 88, no. 2 (August 2011): 471–493.

48 Ambroise Wonkam and Adebowale Adeyemo, "Leveraging Our Common African Origins to Understand Human Evolution and Health," *Cell Genomics* 3, no. 3 (March 2023); Ambroise Wonkam et al., "Five Priorities of African Genomics Research: The Next Frontier," *Annual Review of Genomics and Human Genetics* 23 (2022): 499–521.

Epilogue

1 Noel Ignatiev, *How the Irish Became White* (New York: Routledge, 2012); Karen Brodkin, *How Jews Became White Folks and What That Says About Race in America* (New Jersey: Rutgers University Press, 1998); David R. Roediger, *The Wages of Whiteness: Race and the Making of the American Working Class* (New York: Verso, 2007).

ACKNOWLEDGMENTS

My mother sometimes worried that her daughter might be too timid and bookish to find safety in this world of racial strife. But in time I did, and I am now completing my thirty-first year at the University of North Texas as a Professor of African Evolutionary History. It may not be coincidental that UNT is a vibrant research institution and has one of the top jazz/theater programs in the nation. This intellectual and creative sanctuary has allowed me to explore the diverse ways of knowing with which we as humans have been gifted.

One of the few colleagues with a longer tenure at the university than I, Gustav Seligmann, has been my motivational coach and much more. He tirelessly helped me to sculpt my own personal space within this sometimes bewilderingly large academic community. I am grateful to him and to my colleagues in the History Department at UNT, particularly Jennifer Wallach, chair of the History Department, for their tireless support. It could not have been more fortuitous that Jennifer's inauguration of food studies as an integral part of our curriculum has served to enrich my own understandings of food history. Rachel Moran, a new colleague who specializes in the history of medicine, has also been a help to me in unwinding the sometimes-twisted narratives of health, medicine, and history.

The one individual on this university campus to whom I owe the greatest debt of gratitude is Denise Perry Simmons—biologist, cancer researcher, and senior scientist at the UNT College of Engineering. Since our paths first crossed in 2017, we have spent hours discussing, arguing about, and exploring the practical applications of my research. And, ever so gently, Denise schooled me on ways of conveying my work using the technical symbols of science rather than a style honed from years of communicating in the voice of a history-teaching storyteller.

Given the transdisciplinary nature of this book, I have embraced with gratitude the wealth of advice offered by Harvard-Radcliffe friends and classmates with whom I have remained in touch over the years, including Stephen S. Hill, Myles V. Lynk, Linda Shortliffe, Esther Dyson, Connie Jacowitz, and Alan Weisbard. In those moments of deepest self-doubt and hesitancy, it was the loving support of Marjorie Starkman, MD, and the late biologist Scott Moody, PhD, who saw me through to the finish line.

In reflecting on the narrative thread of this book, I could not help noting the nudge that tipped me beyond the boundaries of my comfort zone and ultimately launched this journey of discovery. A 2007–2008 award from the Fulbright Association allowed me to assume a visiting professorship at the University of the Ryukyus in Nishihara, Japan. I simply cannot find words to adequately express the depth of insights I gained from the professional relationships and friendships that I developed while in Japan.

The emotional support of my husband, Terrill, son, Kenneth, brother, John, and sister, Ida, has been integral to completing this project. I must also include my nephew, John K. Hilliard, who continued to express enthusiasm for this research long past the point that family politeness might have dictated.

My editors at Harvard University Press, Janice Audet, Rachel Field, and Joy Scott Ressler, made the pragmatic difference between having an exuberant jumble of ideas and a coherent book. I simply cannot find words that would convey the depth of my gratitude for their professionalism, patience, and encouragement.

INDEX

Note: Page numbers in italics refer to figures. Numbers in bold indicate a table.

Abd Al-Sadi, 17, *18*
Adeles, Alice, 143
affirmative action, 160
Africa: coastal, 5, 6, 11, 17, 21, 23–29, *26*, 63, 134; East, 8, 134, 140; equatorial, 29; ethnolinguistic groups in, 62; genetic diversity in, 74; history of, 23–28; South, 19, 37, 140; sub-Saharan, 23, 113, 139, 154; West, 5, 7, 11, 15–16, 19, 21, 23–31, 61, 62, 67, 69, 72–74, 81–85, 87, 89, 93–94, 100–104, 112, 116–117, 119–122, 126–127, 132, 134. *See also* Timbuktu; *specific African countries by name*
African Americans: DNA sequences/genome of, 61–63, 78, 80; exploitation of, 57–59; history of, 29; incarceration of, 152–153; misidentification of, 111; "passing" for White, 37, 89, 116–117; risks during pregnancy, 96–97; of slave descent, 6, 8, 10, 53, 61, 63, 72, 79, 89, 104, 112, 114, 116, 121–122, 126, 132, 135, 139, 144, 145; vitamin-D deficient, 146
African and African American Studies, 160, 161
African genome, 41–42, 50–51, 62, 67, 116, 118, 156, 161
African Genome Project, 154
African sleeping sickness, 31
African Society of Human Genetics, 156
Africanus, Leo (Al-Hasan Muhammad al-Wazzan), 15
Afrikaners, 37
Afro-Caribbeans, 63, 74, 82, 153
Agassiz, Louis, 45–46
Akan people, 23, 25
Al-Bakri, Ubaidalla, 17
Alfonso I (king of Kongo), 24
algorithms: ENP Model, 121–124, 127, 128; and health care, 111–112; for kidney transplants, 114, 116; race-correction, 112–113; vaginal birth after cesarean (VBAC), 113
alleles, 45, 54, 65, 79, 133. *See also* Single Nucleotide Polymorphisms (SNPs)
All of Us project, 155
Al-Mas'udi, 21
Aloia, John, 120
Alzheimer's disease, 135–136
American Anthropological Association, 45
American Association for Cancer Research (AACR), 87
American Heart Association, 107
American Society of Nephrology, 119

Angiotensin-converting enzyme (ACE) inhibitors, 142–143
Angola, 100–101, 159, 162
anthropology, physical, 69
anti-eugenics, 66
Apolipoprotein L1 (*APOL1*) gene, 11, 30–31, 63, 65, 115, 116, 121–123, 145
Ashkenazi Jews, 52
Asians, 8, 37, 45, 54, 72, 116, 130, 136, 145, 159
Association of Black Cardiologists, 141, 142
asthma, 79
atrial fibrillation, 64
azathioprine, 143

Bell Curve, The, 71–72
Big Data, 41, 63, 86, 111
biochemistry, 155
biogeographical ancestry, 49, 120
bioinformatics, 155
biopiracy, 59–61
blood pressure disorders, 96, 97. *See also* hypertesion
BoneKEy Reports, 85
bone mineral density (BMD), 83–84
BRCA (BReast CAncer gene), 90
Broad Institute, 64
Bryc, Katarzyna, 116
Burgoyne, Alexander P., 77, 78

calcium: recommended daily dose, 89, 91; renal retention of, 88–89
calcium absorption/retention, 83, 93
calcium deficiency, 94, 148
calcium intake, 89, 90, 91, 96, 121, 123, 124, 127; and lactose intolerance, 100; during pregnancy, 93–95, 98; and *TRPV6*-expressing cancers, 147–150
calcium ion channels, 62, 63, 84–85, 121, 122; channel-blocking drugs, 148–150
calcium metabolism, 98
cancer: breast, 58, 64, 82, 90, 136, 149; cervical, 90; chemotherapy-resistant, 91; colorectal, 57, 147, 148; DNA sequences for, 80; and the HGP, 35; high risk of, 4, 147, 149; and hypercalcemia, 84; metastatic prostate, 11, 85–86, *86*, 147, 148; microbiome links to, 109; ovarian, 137, 138, 147, 148; pancreatic, 98; prostate, 57–58, 79, 136, 149; research on, 87; triple negative breast (TNBC), 11, 85–86, *86*, 90, 90–91, 140, 145, 147, 148; *TRPV6*-expressing, 84–92, 121, 147–150
cancer genomics, 91
capsaicin, 150
cardiometabolic disease, 79
cardiomyopathy, 96
cardiovascular disease/disorders, 14, 32, 33, 63, 64, 97, 108, 113, 136, 142, 144, 145; heart failure, 32, 141–142, 144
Carney, Judith, 102–103
Catalan Atlas, *22*
cell-line errors, 57
Centers for Disease Control and Prevention (CDC), 13, 96
Chavira, Aries, 109
chemotherapy, 53, 90, 91
childbirth mortality, 4, 11, 95, 97, 113

cholesterol, 144
Chongqing Medical University, 149
civil rights movement, 105
climate change, 49–50
cognitive performance, 66
colonialism, 42
Columbus, Christopher, 23
Common Disease, Common Variant (CDCV) hypothesis, 54, 55
congestive heart failure. *See* cardiovascular disease/disorders
Congressional Black Caucus, 141
COVID-19, 14
craniometry, 69
creatinine, 114, 116, 118, 128
Crick, Francis, 43
critical race theory, 160
Crowley, Jordan, 113–114, 117, 118
Cutler, David, 29
cystic fibrosis, 52
cytotoxicity, 149

Darwin, Charles, 69–70
Database of Genotypes and Phenotypes (dbGaP), 154
Dawson, Charles, 70
Debes, Jose D., 149
dementia, 135–136
democracy, 158–159, 160
Denisovans, 40
DeSalle, Rob, 117
diabetes: gestational, 136; type 2, 11, 13, 64, 79, 97, 104, 108, 136, 144, 147
diet: gluten-free, 102; high-calcium, 8; high-sodium, 6, 32–33, 116; low-sodium, 5, 11, 16–17, 28, 101–102, 121, 140, 141, 144, 146; soul food, 105–106; transition from rice and corn to wheat, 104–105. *See also* lactose tolerance/intolerance; nutritional guidelines
Dietary Approaches to Stop Hypertension (DASH), 108
dietary diversity (DD), 106–107
Dietary Guidelines for Americans, 107–108
DiNicolantonio, James, 32
Diodorus Siculus, 15
disease: agricultural, 103; brought to the New World, 23; chronic, 106, 107, 108; neurological, 109; shared risk of, 123; susceptibility to, 51; tropical, 103
DNA ancestry, 49, 52, 60–61, 116, 119–120, 122, 132, 137
DNA sequences, 35, 43; of African Americans, 80, 104; for cancer genomes, 53, 80; and intelligence, 77; Neanderthal, 39; West African, 83–84
double-helix model, 43
drug metabolism, 109

eclampsia, 96
Ecological Niche Populations (ENPs), 10, 11, 112, 131–134; African American/Sodium-Metabolic (Disparities), 121; components of, *123*; defining, 122–124; identification of, 140; West African/calcium, 121; West African/sodium, 121
Ecological Niche Populations (ENPs) Model, 120–126, 132, 143; applying, **124–126**; caveats, 127–129

ecological niches, 9
educational achievement, 66, 67
Eighth Joint National Committee, 142
end-stage renal disease (ESRD), 30, 145. *See also* kidney disease
enterotypes, 109
epigenetics, 78
Epstein, David, 73
essentialism, 51, 68, 156; racial, 39–40, 42, 44, *44*
Estimated Glomerular Filtration Rate (eGFR), 6, 114
ether use, 59
ethnic cleansing, 71
ethnolinguistic groups, 60, 62, 63, 74, 159–160
ethno-pharmacogenomics, 140–143
ethnopharmacology, 128
eugenics, 11, 39, *44*, 46, 66, 68–75, 78, 153–154, 156
Eurocentrism, 11
evolution, 68, 78
evolutionary biology, 132–133

Florida International University, 124
Floyd, George, 103
focal segmental glomerulosclerosis (FSGS), 30
Francis-Lyon, Patricia, 87
Fryer, Roland G., 29

Galton, Francis, 69
Gates, Henry Louis, Jr., 16
genetic adaptations, 4, 6, 8, 9–11, 21, 101, 120–121, 123, 146
genetic ancestry, 6, 31, 39, 57, 60, 64, 67, 96, 111, 116–118, 134
genetic architecture, 78–79, 131
genetic diversity, 10, *38*, 42, 62, 130, 134, 154, 156; of Africans, 36–37, 39, 62, 113, 134–135; devaluation of, 36–43; and the Lewontin Paradox, 45–49; nested subsets, 43–45; stratification dilemma, 50–57
genetic research, 80, 109–110, 117, 154, 155
genetics, 8, 41, 68, 132, 134, 155, 156; medical, 139; population, 46
gene variants, 43, 54, 63–64, 133, 136; in Africa, 55–56, 74; Apolipoprotein L1 (*APOL1*) gene, 11, 30–31, 63, 65, 115, 116, 121–123, 145; *BRCA*, 90; in China, 55; and genetic diversity, 134; identification of, 65–66; linked to hypertrophic cardiomyopathy, 139; population-specific, 55, 109; potential disease-causative, 52, 54, 56, 60, 61–63, 66, 141; shared, 55; *TRPV6*, 93–94, 98, 121, 122, 147–150
genome: African, 41–42, 50–51, 62, 67, 116, 118, 156, 161; African American, 62; Danish, 62; European, 56; Irish, 62
Genome-Wide Association Studies (GWAS), 41, 52–53, 54, 57, 61, 62, 64, 65, 67, 79, 80, 88, 130, 155, 156
genomic medicine, 11, 128, 145
genomics, 8, 41, 52, 54, 55, 68, 80, 108, 128, 133, 155; and African Americans, 57, 61–63; ancestral, 130–157, 133; cancer, 91; and Ecological Niche Populations (ENPs), 131–134
genotypes: African, 37, 147; ancestral, 88; differences in, 6, 115, 127; European, 38, 77, 92; identification of, 56, 78, 79, 154

geo-bio-climatic selection, 78
gestational diabetes, 136
Ghana, 5, 11, 16, 17, 18, 20, 21, 139
Glazer, Nathan, 29
global health, 154, 156
gold trade, 4–5, 11, 15–21, 23, 25
Graves, Joseph L., 69, 132–133
Great Migration, 104
Grim, Clarence, 29
Gullah Geechee communities, 74
Gupta, Vinod K., 109
gynecology, 58–59

hair relaxers, 137–138
Hambrick, David Z., 77
Hannah-Shmouni, Fady, 31
haplogroups, 130
haplotypes: sharing, 39–40; *TRPV6*b, 85, 88
Harden, Kathryn Paige, 66
Harvard University, 16, 46, 67, 160, 161
healthcare inequities, 112
health disparities, 8, 12, 134, 135, 139; ancestral etiology of, 120; gap in, 62; racial and ethnic, 31, 147; social determinants, 122
health equity, 92
heart disease. *See* cardiovascular disease/disorders
Hemings, Sally, 69
hereditarian hypothesis, 66
Herodotus, 15, 20–21
Herrnstein, Richard, 71
heterosis, 71, 74
high blood pressure. *See* hypertension
Hispanics. *See* Latinos
Hitler, Adolf, 71
Holwerda, Minke, 49–50
homogeneity, 55, 56, 60, 62, 63, 134
Hooker, Stanley E., Jr., 57
Hsu, Joy, 118
human exceptionalism, 74, 75
Human Genome Project (HGP), 7, 34, 35, 45, 52, 68, 136
Human Heredity and Health in Africa (H3Africa) Consortium, 154, 156
Human Pangenome Reference Consortium (HPRC), 41, 43
Human Pangenome Reference Sequence Project, 155
hybrid vigor, 71, 74
hypercalcemia, 84
hypertension: in African Americans, 4, 6, 11, 13–14, 29–31, 34, 63, 79, 96, 101, 108, 112, 121, 122, 135, 140, 142, 144, 145, 147; and calcium, 98; chronic, 96; and diet, 108; genetic diversity, 135; gestational, 96, 99; salt-sensitive, 28–31, 98–99, 144; triggers for, 30–31, 136; and vitamin D, 147
hypertrophic cardiomyopathy, 139

Ibn Khaldun, Abd al-Rahman, 15
Ibn Khurdadhbih, 21
immunosuppressive medication, 143
immunotherapy, 53
indigenous societies, 19, 23, 41, 42, 47, 102, 154. *See also* Native Americans
infant mortality, 95
inflammatory bowel disease/syndrome, 64, 79
Intelligence Quotients (IQ), 67, 70, 72, 76, 77–78
International HapMap Project, 155

Irving, Shalon, 97
isosorbide dinitrate/hydralazine (BiDil), 141–142

Jefferson, Thomas, 69
Jensen, Arthur, 51, 72
João III (king of Portugal), 24–25
Johns Hopkins University, 57
Jordan, David Starr, 71
Juettner, Otto, 133
Jurashcek, Stephen, 108

Kahn, Jonathan, 141, 142
Kenya, 42, 56, 139, 156
Khera, Amit V., 64
Khoisan people, 37–38, 40, 47. *See also* San people
kidney disease, 11, 30, 79, 114, 116, 145
kidney failure, 1–3, 4, 7, 13, 14, 28, 31, 34, 63, 121, 122, 144
kidney stones, 113
kidney transplants, 113–115
kidney transplant score, 124
Kittles, Rick, 45
Knowles, Beyoncé, 97
Kongo, kingdom of, 24
Kuhn, Thomas, 99

lactase-phlorizin hydrolase (*LCT*) gene, 55–56
lactose tolerance/intolerance, 55, 56, 83, 93, 94, 95, 99–100, 121, 140, 148
Las Casas, Bartolomé de, 23
Latinos, 8, 41, 59, 145, 153, 159
Lewontin, Richard, 46
Lewontin Paradox, 45–49, 50
Lichtenstadter, Ilse, 14–15
lidocaine, 150
Lombardo, Paul A., 71
Long, Jeffrey, 44–45
Lupien, Mathieu, 91
Lynn, Richard, 76

machine learning, 111
malaria, 51, 140
Mali, 4–5, 11, 16, 17, 20, 21, 26
malnutrition, 58
Mankind Quarterly, 78
Mansa Musa, 4, 21, *22*, 23
Martin, Alicia, 64
Massachusetts General Hospital, 64
maternal mortality, 4, 11, 95, 97, 113
Matthews, John, 27
Mbemba, Nzinga, 24
medical capitalism, 80
medical genetics, 139
medical histories, 106
medical research, 9–10
medical training, 153
medicine: genomic, 11, 128, 145; in Japan, 1–2, 3, 6, 115–116, 128; precision, 53, 57, 60, 79, 80, 112, 113, 115–120, 128; racialized, 7, 52, 114, 133, 141, 142
metagenomics, 78, 108
microbiome, 108–109
Middle Passage, 29
misdiagnoses, 1–3, 112
molecular biology, 68
Morton, Samuel George, 69
Murray, Charles, 71
Museum of Comparative Zoology, 45–46

nagana, 82
National Association for the Advancement of Colored People (NAACP), 141–142
National Basketball Association (NBA), 72
National High Blood Pressure Education Program Working Group on High Blood Pressure in Pregnancy, 96
National Human Genome Research Institute (NHGRI), 137
National Institutes of Health (NIH), 138, 154, 155
National Kidney Foundation, 119
National Medical Association, 141
National Science Foundation, 152
Native Americans, 8, 23, 59–61, 103, 112, 121–123, 130, 145, 153. *See also* indigenous societies
natural immunity, 47, 50, 51, 103
natural selection, 66
Nazi Germany, 70–71
Neanderthals, 39
nested subsets, 43–45, *44*
New York University Langone Health, 120
Nigeria, 104, 139, 155
nonalcoholic fatty liver disease (NAFLD), 136
Northwestern University, 98
nuclear factor of activated T-cells (NFAT), 149
nucleotides: base pairs, 35; diversity, 37, 45. *See also* Single Nucleotide Polymorphisms (SNPs)
nutritional guidelines, 32–34, 89, 143–145, 148. *See also* diet

Obama, Barack, 29
Obasogie, David E., 150
obesity, 104, 109
obstetrics, 93–95, 97
Olopade, Olufunmilayo, 91
oncology, precision, 53. *See also* cancer
1000 Genomes Project, 154–155
organ transplantation, 3, 7, 113–115
osteoporosis, 8, 83, 85, 99, 148

pangenomes, 41, 43
Pan-Human Genome, 155
Park, Mungo, 27
partial randomness, 127
pastoralism, 8, 93
Paul III (pope), 23
pharmacogenomics, 91. *See also* ethno-pharmacogenomics
pharmacology, 53, 143, 155; and cancer, 35; reactions to, 140
phenotypes, 55, 67; Eurasian, 49; racial, 40
Piltdown Man, 70
Polygenic Risk Scores (PRSs), 63–68
polygenic risk studies, 62
polymorphisms, 134; non-European, 84. *See also* Single Nucleotide Polymorphisms (SNPs)
population genetics, 46
population resilience, 48
Portland State University, 39
poverty, 8, 42, 122

precision medicine, 53, 57, 60, 79, 80, 112, 113, 128; and the colorblindness trap, 115–120
preeclampsia, 96, 97, 98
pregnancy, 92–99; and maternal mortality, 4, 11, 95, 97, 113
Princess Margaret Cancer Centre (Toronto), 91
proteomics, 108

quality controls, 135–140

race: biological distinctions, 51; designations of, 8–9; and disease risk, 3; erroneous assumptions about, 9; and eugenics, 68–75; and medical diagnosis, 2; and self-identification, 159
race-free calculator, 119
racial classification fallacy, 139
racial discrimination, 8, 30, 115, 122
racialized medicine, 7, 52, 114, 133, 141, 142
racial superiority, 44, 71
racism, 24, 76, 95, 109, 122, 134; scientific, 51, 78; and stress, 31, 79, 96, 97; systemic, 30
radiation therapy, 91
Rainforest Hunter-Gatherer (RHG) groups, 154
religion, 81–82
renal failure. *See* kidney failure
renin-antiotensin-aldosterone system (RAAS), 30
renin-angiotensin system (RAS), 30
Resnick, Brian, 53
Resnick, Lawrence Malcolm, 98
rice, 102–105
Richards, Todd, 105
RNA transcripts, 108
Roberts, Dorothy, 117
Rotimi, Charles N., 155
Rush, Benjamin, 69
Rushton, Philippe, 71
ruthenium red, 150

Saini, Angela, 76
salt, 5, 17–19, 23, 27; historic uses of, 28–29; sensitivity to, 99, 145. *See also* sodium intake
Salzberg, Steven, 57
Sangumba, Jorge, 158–159
San people, 137, 154. *See also* Khoisan people
Sarich, Vincent, 72
Saulsberry, Loren, 91
science, racist, 51, 78
science education, 150–153
segregation, 158
self-care, preventive, 30
Senegal, 92–93
Senegambia, 61, 102–103
sexual exploitation, 73–74, 114
Shockley, William, 51
sickle cell anemia, 51–52, 60–61, 119, 131, 140
silent barter, 20–21
Sims, J. Marion, 58–59
Single Nucleotide Polymorphisms (SNPs), 45, 53, 54, 77
skull shapes and sizes, 39–40, 75
slavery: Black Americans descended from, 6, 8, 10, 53, 61, 63, 72, 79, 89, 104, 112, 114, 116, 121–122, 126, 132,

135, 139, 144, 145; history of, 5, 16–17, 23–28, 61–62, 69, 87, 103, 160; legacy of, 71, 73, 109, 152; and medical science, 58; and sexual exploitation, 73–74. *See also* African Americans, of slave descent; trans-Atlantic slave trade
Slave Ship Hypothesis, 29
social inequality, 66
socioeconomic status, 66, 78, 97, 151
sodium intake, 5, 14, 96, 101–102, 116, 124, 127, 128, 143–145; during pregnancy, 98; gain-of-function variants, 31; normal, 30; recommended, 32–34. *See also* salt
sodium toxicity, 32
Songhay, 5, 11, 16, 17, 20, 21
SOR-C13, 150
Soricimed BioPharma, 149
soul food, 105–106
South Africa, 37–38
Spanish conquistadores, 23
spinal muscular atrophy (SMA), 52
Sports Gene, The, 73
Stanford University, 67, 71
STEM education, 151–153
Stewart, Jack, 149–150
STONE score, 113
Strabo, 21
Stratakis, Constantine A., 31
stratification dilemma, 50–57
Strauss, William L., 70
stress: effects of, 30, 31, 79; related to racism, 31, 79, 96, 97
stroke, 13–14, 32, 136
Sublimus Deus (papal bull), 23
subsistence farming, 16, 19–20
Sudan, 23, 55, 55–56
Sufism, 15
syphilis, 59

T2T Consortium, 41
Tanzania, 42, 56
Tārīkh al-Sūdān, 17, *18*
Tate, Florence, 13
Tay-Sachs, 52
Terry, Bryant, 101
tetany, 58
Timbuktu, 4–5, 11, 15–18, 20, 21, 76, 158
Timmons, Greg, 103
Tishkoff, Sarah, 55, 117
Torres, Jada Benn, 45
toxemia, 97
trans-Atlantic slave trade, 25–28, 51, 63, 103–104. *See also* slavery
transcriptomics, 108
Trans-National Institutes of Health (Trans-NIH), 155
trans-Saharan trade, 23
trauma, historical, 132
triglycerides, 144
Trypanosoma brucei, 31, 82, 140, 145
Trypanosomiasis, 31, 145
Tsetse Belt, 7, 82–84, 120–121, 140, 148
tsetse fly, 19, 31, 82
Tucker, William H., 71
Turing, Alan, 111
Tuskegee Syphilis Experiment, 59
type 2 diabetes, 11, 13, 64, 79, 97, 104, 108, 136, 144, 147

"Uncle Ben," 103
University of California, Berkeley, 72

University of California San Francisco, 143
University of Chicago Department of Ecology and Evolution, 37
University of London School of Oriental and African Studies, 158
University of New Mexico Anthropology Department, 44–45
University of North Texas, 104
University of Pennsylvania Perelman School of Medicine, 55, 119
University of San Francisco Health Informatics Program, 87
University of Virginia, 71
University of Wisconsin, 39
urology, 113
US Air Force Academy, 60
US Department of Agriculture (USDA), 32, 100, 107, 123, 145
US Department of Health and Human Services (HHS), 107
US Food and Drug Administration (FDA), 141
US National Academies of Sciences, Engineering and Medicine, 117
uterine fibroids, 137, 138

vaccines, 47
vaginal birth after cesarean (VBAC), 113
Vanhanen, Tatu, 76
vesicovaginal fistulas, 58–59
vitamin D: deficiency, 146–147; synthesis, 40
Voting Rights Act, 159

Wallach, Jennifer, 104–105
Wang, Xuexia, 124
Washington, Booker T., 105
Washington, Harriet, 58–59
Watson, James, 43
Wayne State University Medical Center, 98
Weasel, Lisa, 40
Weill-Cornell Medical Center, 98
White Men Can't Jump (film), 72
White supremacy, 71, 158
Williams, Serena, 97
Willis, Virginia, 105–106
Wilson, Thomas W., 29
Winickoff, David E., 150
Woodley, Michael, 78
World Health Organization (WHO), 127

X-ray exposure, 133

Yancy, Clyde W., 141–142
Yanomani Indians, 30
Yu, Ning, 37
Yudell, Michael, 117

Zilbermint, Mihail, 31
Zimmerman, K. D., 74